All Too Human

Understanding and Improving Our Relationships with Technology

Anne McLaughlin
North Carolina State University

CAMBRIDGE
UNIVERSITY PRESS

University Printing House, Cambridge CB2 8BS, United Kingdom

One Liberty Plaza, 20th Floor, New York, NY 10006, USA

477 Williamstown Road, Port Melbourne, VIC 3207, Australia

314–321, 3rd Floor, Plot 3, Splendor Forum, Jasola District Centre, New Delhi – 110025, India

103 Penang Road, #05–06/07, Visioncrest Commercial, Singapore 238467

Cambridge University Press is part of the University of Cambridge.

It furthers the University's mission by disseminating knowledge in the pursuit of education, learning, and research at the highest international levels of excellence.

www.cambridge.org
Information on this title: www.cambridge.org/9781316515600
DOI: 10.1017/9781009026093

First published 2022

A catalogue record for this publication is available from the British Library.

Library of Congress Cataloging-in-Publication Data
NAMES: McLaughlin, Anne, 1976- author.
TITLE: All too human : understanding and improving our relationships with technology / Anne McLaughlin, North Carolina State University.
DESCRIPTION: Cambridge, United Kingdom ; New York, NY, USA : Cambridge University Press, 2021. | Includes bibliographical references and index.
IDENTIFIERS: LCCN 2021035074 (print) | LCCN 2021035075 (ebook) | ISBN 9781316515600 (hardback) | ISBN 9781009012546 (paperback) | ISBN 9781009026093 (epub)
SUBJECTS: LCSH: Technology–Social aspects. | Technology–Psychological aspects.
Classification: LCC T14.5 .M33 2021 (print) | LCC T14.5 (ebook) | DDC 303.48/3–dc23
LC record available at https://lccn.loc.gov/2021035074
LC ebook record available at https://lccn.loc.gov/2021035075

ISBN 978-1-316-51560-0 Hardback
ISBN 978-1-009-01254-6 Paperback

CONTENTS

FIGURES

TABLES

PREFACE

At parties I am often asked what I do. The career "professor" makes sense to everyone, so I start with that. But when I say I am a psychologist, the next question is often, "Are you reading my mind?" Yes, I want to say, I knew you would ask me that. Everyone does. Instead, I explain that I'm a human factors psychologist, one who studies some of humans' greatest feats and foibles in the hopes of building a better world. It's a tough elevator pitch, but one I'm passionate about. I truly believe that the more people know about *people*, the more they will demand that technology and systems be built well, rather than bending over backwards to accommodate poor design and then blaming themselves when they fail. By the end of this book, I hope this is the position readers will take. Armed with an understanding of the basics of our all-too-human brains and bodies, they will know we can't be expected to be superhuman. And yet, we can *seem* superhuman when the world around us is suitably designed.

Come with me for an adventure that spans the Earth and time, from the depths of the Chernobyl nuclear reactor to airplanes in the skies above. From online dating to getting naked in Washington, DC. Discover the link between prisons in the desert and how a pandemic can spread beyond control. Learn how to harness evolution to save lives on the road at night. These are just a few of the examples brought together to illustrate the capabilities and limitations of the human race. Each example draws from amazing discoveries about the mind, and then goes beyond them. I will show you how to use this knowledge to explain some of our human troubles and prevent them in the future. Each chapter focuses on an aspect of human cognition: attention, memory, creativity, problem-solving, decision-making, or hard-wired biases, illustrated with real-word examples and events. What keeps this from being a book on cognitive psychology is that each chapter also focuses on ways humans have (or could) change products, systems, and their surroundings to pair with human cognition.

Officially, I'm a psychologist. But I wrote this book because I am a humanist, a technologist, and an optimist. I can appreciate all that technology and society does for me, even while it is a source of enduring frustration. With the knowledge and tools in this book we can join together to prevent that aggravation and wasted time. We can make the world a safer, more efficient, and more enjoyable place. We can better decide when technology will improve or decrease our quality of life. We can help the professionals in our lives such as doctors, lawyers, and politicians make better decisions for us and also make better decisions for ourselves.

ACKNOWLEDGMENTS

I am forever grateful to my editor, Kate Davis Jones. Kate, you kept me on target for a year, tirelessly helping me to clarify and extend my examples, all the while pushing me to never write the words "for example." Your style is immaculate and I am thankful for your help in tightening my own.

As might be expected from an author so interested in usability, I want to give many thanks to the people who "tested" my chapters and illustrations for readability and enjoyment. I couldn't have written (and rewritten and rewritten) this without: Amy Greene, Claudia Lee, Rachel Thiel, Emily Ford, Kendyl James, Tanvi Thummar, Mitch McDonald, Imani Murph, and the students from my graduate courses in Human Factors Psychology at North Carolina State University. Go Pack!

Thank you to my colleagues who vetted my chapters for accuracy, graciously giving me their time and expertise. Your work is the backbone on which we will improve the world: Lynne Baker-Ward, Rick Tyrrell, Chris Wickens, Maribeth Gandy Coleman, Laura Levy, Frank Drews, Arathi Sethumadhavan, and Ericka Rovira. Thank you to Matt Shipman for giving me the encouragement to pursue this idea as a book, and to Cat Warren for mentoring me in how to find a publisher. I value and appreciate you all. I'd also like to thank my dog Royal for his willingness to serve as an animal example throughout the book and for forcing me to get up occasionally to let him in or out of the house.

Lastly, I want to give thanks to my family. To my dad, who put in a great amount of effort to understand what an "Engineering Psychologist" was and why I'd want to be one. To my mom, who is the most creative problem-solver I've ever known and always a good illustration of how people will bend systems to their will, whether the designer intended that or not. Most of all, I owe thanks to my husband, Tom. You were there for long walks and endless discussions of examples I wanted to include. You helped me to focus and prioritize. Even after a long day, you were always willing to read through my work and ask the right questions. I know I can count on you and you always lift me higher, in this and in everything. You're the greatest.

INTRODUCTION

> To understand the actual world as it is, not as we should wish it to be, is the beginning of wisdom.
>
> – Bertrand Russell, 1934, *Mortals and Others*, V. II

Sometimes designs are so bad, they hit you in the face.

In 1993, I was a 16-year-old Alabama driver with a new red Saturn. That Saturn gave me independence, and I loved that car from its pointy hood to its strange chopped-off rear. It was full of fancy new technology, including non-denting plastic doors and a motorized seat belt connected to a track along the door frame. I'm not sure why an automatic shoulder belt was a selling point, since you still had to buckle the lap belt, but it felt like the future.

In lower Alabama (or, as the locals call it, L.A.), if it's not dark clouds pouring torrential rain, it's a white-hot sun that turns a black steering wheel into a branding iron. I wielded the car's visor to battle the blinding sun: down in front, rotate left, rotate back, wherever the sun shone. It was a constant dance to make driving bearable.

When the Saturn attacked me, I had the visor down and rotated left, shielding the driver's side window. I parked, and when I opened the door to exit, *BAM!* Something slapped me in the face. I was dazed, and my first reaction was shock followed by raising my hands to protect myself. But no one was there. As my vision cleared I took stock of the damage.

My nose was scratched and I had a welt forming on my left cheek. I looked around for a tree branch or a baseball that could have hit me through the open door. It took some time but I identified the villain: the shoulder belt that was now nestled tidily into the right side of the driver's seat. It had disconnected from the door and slapped me across the face as it retracted. The heavy buckle hit me on the way down, giving me the black eye.

I mentally rebuilt the events. When I opened the door, the automatic seat belt tried to move forward on the track and encountered the sun visor. Because the release button was on the forward-facing side of the shoulder belt, when it hit the visor it disconnected, whipping down across my face. The visor wasn't huge, and it wasn't in the way of getting out of the car. I hadn't even thought about moving it before opening the door.

I wouldn't hear the words "human factors" for another four years, but in that moment I started thinking like a human factors psychologist. Why was the release button on the front of the shoulder belt? Who could possibly remember to move the sun visor to the front before opening the door? Why didn't the designers see this coming? Why was I so embarrassed that this happened to me?

I'd like to say that was the only time I didn't move the sun visor back before getting out of the car, but of course it wasn't. In the six years I drove it, I was smacked by the heavy plastic buckle about ten times. Each time I had that same reaction: "I can't believe I forgot *again*."

We Do It to Ourselves

I'm certainly all too human. We all are. We get fooled again and again, whether it's trying to push a door that says pull or to remember why we walked into the kitchen. The importance of the task is almost immaterial – it's as easy to forget to pick up milk as it is to forget a dog in the backseat of a hot car. Indeed, the penalty for forgetting to move the sun visor was high – ask my poor nose – but I couldn't remember to move the visor back from the side window before opening the car door. Our human brains aren't capable of remembering what we don't remember or constantly paying attention. Our constant (and predictable) failures are as central to our humanity as love, ambition, or any number of positive characteristics. But what can we do to help our all-too-human selves?

The first step in stopping mistakes is to acknowledge that making them is often beyond our control. Thanks to millions of years of evolution our minds react in predictable, if sometimes undesirable, ways to our surroundings. The second step is to understand consciousness and behavior at a level where we can predict how people will act. Third is to use this to create the world around us, working with our talents and acknowledging our failings. Once we understand how people see, hear, feel, and think we can make a world that protects us rather than "fools" us. The answer wasn't for me to remember to put the sun visor back, it was to change the situation so I didn't *need* to remember. But this kind of thinking, where we make the world around us easier, is more recent than most people realize.

As technology progressed during the second Industrial Revolution, machine development quickly outpaced our ability to adapt. Once our lives started to depend on these technologies, we needed a translator between the person and the machine: an interface. Before this, when someone could only move as fast as their feet could run, or only needed to dodge something as fast as it could be thrown, there was not much need to worry about how we used machines. For example, a gas pedal interprets the will of the driver regarding speed and the steering wheel interprets how they want the car to turn. A dial interprets which gas burner to ignite and how high. The phone dial pad or contact list interprets whom we wish to call. All of this began right at the turn of the twentieth century.

Thus, human-centered design began in the second Industrial Revolution, after years of machine-related deaths and crippled workers in factories. Some of the first people to recognize the disconnect between work demands and human capabilities were Frank and Lillian Gilbreth. Together, they founded a new field, one that took into account the person doing the work as well as the machine being worked with. The Gilbreths engaged with the same questions that confound modern designers. Today we worry about controlling autonomous cars. The Gilbreths had to understand how the public would react to *regular* cars, which moved a lot faster than a horse.

The Gilbreths (famously profiled with their many children in the book and film *Cheaper by the Dozen*) filmed people while working. They called these "time-motion studies." From these films they zeroed in on how to make motions smaller, more efficient, and with better flow. They proposed surgeons keep their tools on a tray, arranged according to the most frequently used and the order in which they would likely be called for during surgery. Another was to build a raised stand for bricklayers, so that the worker didn't bend over to grab every single brick. They even enrolled the family in their research, as remembered by their son Frank, Jr., and daughter Ernestine:

> "Is it better to stack the dishes on the table, so that you can carry out a big pile?" Dad asked. "Or is it better to take a few of them at a time into the butler's pantry, where you can rinse them while you stack. After dinner we'll divide the table into two parts, and try one method on one part and the other method on the other. I'll time you."[1]

The Gilbreths were true empiricists. Optimizing motion may seem to have obvious benefits, but there was pushback from naysayers who claimed that lazy employees should simply work harder, and that the promise of increased production in shorter time *with easier work* was a dream. The Gilbreths were undeterred. Lillian Gilbreth even counted the "happiness minutes" of the worker as an important measure of work. She brought the radical idea that our world should be built to fit us, mentally and physically, rather than forcing the worker to adapt to whatever poorly engineered system was put in front of them – and all of this in a time when the announcement of her engagement noted that "Although a graduate of the University of California the bride is nonetheless an extremely attractive young woman."[2] That graduate deserves credit for being the first to consider the "human factor" in work, setting up the criteria that would be used by psychologists in the World War II.

Enter the Human Factor

Frank Gilbreth might have liked the controls used in the planes of World War II as they were clustered closely together, meaning smaller movements were needed to use them. But as aviators in World War II found out, this resulted in pilots accidentally retracting the landing gear when they meant to pull up the wing flaps. In a short time during World War II, dozens of planes crashed due mixing up the flaps and landing gear.[3] When your profession has a nickname for an error (in this case, the "Gear Up Club"), you might begin to suspect that there is a problem with the design of the controls. Most drivers have been in a similar club when renting an unfamiliar a car and turning on the wiper instead of the lights, but with less egregious consequences. The thinking of the 1940s was that good pilots don't make errors. If a pilot forgot to put down the landing gear, it meant the wrong pilot had been chosen for the job. But the crashes couldn't be dismissed. The pilots knew how to fly. They weren't suicidal. The planes didn't malfunction. Why did they keep making the same mistake over and over again?

Psychologist Paul Fitts and Capt. Richard Jones decided to look at the problem from the pilot's perspective, asking over 500 World War II pilots to "Describe in detail an error in the operation of a cockpit control which was made by yourself or by another person whom you were watching at the time."[4] As someone who has sent out many surveys only to get a few returned, I wonder if Fitts anticipated the onslaught of comments he would get from pilots – all highly detailed and with many frightening close calls. My favorite was his description of a harrowing incident where the passengers were instructed to "jettison their baggage" to keep the plane aloft. But after saying goodbye to everyone's bags, the pilot realized the troubles came from him forgetting to switch power to an engine. This gave me a new perspective on losing my luggage.

By 1947, Fitts and Jones had finished their report. It was shocking. To get an idea of what pilots were facing, imagine renting a car with the brake and accelerator reversed or in entirely different locations. That's what switching between planes was like. Pilots moved between aircraft that positioned knobs differently for three critical controls: throttle, propeller, and air–fuel mixture. Three types of aircraft, three different locations for these controls. Pilots were expected to learn and adapt. Only often they couldn't, or slipped into a habit formed by flying another type of plane, meaning the more experienced pilots were the most likely to err. Almost all crashes blamed "pilot error," not the control configuration. Here, the pilots were ahead of the engineers – sometimes they glued cardboard triangles to the flaps control and a piece of tire to the landing gear, helping their hands realize when the wrong control was grabbed before they used it.[5]

Rebecca Cameron recounted an example of the entrenched thinking of the time in *Training to fly: Military flight training 1907–1945*. Due to the type of rubber tires on the planes, pilots tended to bounce when landing. Though not fatal, these bouncy landings were costly. When the damage was tallied, the colonel in charge demanded to know "What's the reason for all these broken landing gears? All these broken wings?" When he was told it was "bad landings," he decreed, "Take a memorandum. There will be no more bad landings at this field."[6] You can imagine how well that worked. Such is the all-too-human history of believing that desire and effort can overcome design flaws.

I like to think of it as the difference between prescriptive and descriptive behavior. Prescriptive behavior is what one *should* do. Telling the pilots to land smoothly or use the "correct" controls is prescriptive. Descriptive behavior is what one *does* – pilots land poorly on solid rubber tires. Badly designed tires or confusing controls have predictable results. A good manager or designer recognizes that we have to work with humans as they are, not as we wish them to be. We have to describe how people act, understand them, predict them, and only then can we hope to come up with ways to encourage correct behavior or at least lessen the consequences for a mistake. Those who refuse to understand bad designs are doomed to repeat them.

A Prescription for Description

The gulf between the prescriptive and the descriptive describes many of our issues in living in a human-designed world. But to find solutions we have to understand a multitude of facts about the human mind and body. Researchers have amassed a great deal of this knowledge in cognitive psychology, social psychology, biological psychology, biology, neuroscience, and genetics, but it can be difficult to find an overview of how all those can be applied *right now* to the world around us. That's what I want to share in this book – the connections between these many areas of research with stories "ripped from the headlines." That's a lot of ground to cover, but by the last chapter you will understand enough about human capability and limitation to have unique insights into high-profile news stories and explanations for everyday frustrations and successes. Be warned – once you have an understanding of why we do what we do, you'll start to see the human factor everywhere. Hopefully it won't have to hit you in the face.

1 "BRACE FOR IMPACT"

> One essential characteristic of modern life is that we all depend on systems – on assemblages of people or technologies or both – and among our most profound difficulties is making them work.
>
> – Atul Gawande, 2010, *The Checklist Manifesto: How to Get Things Right*

On a cold January afternoon in 2009, US Airways Flight 1549 departed LaGuardia airport in New York, headed for Charlotte, North Carolina. There had been a light dusting of snow that morning but even with cloudy skies the scenery was beautiful enough along the Hudson River for Captain Chelsey "Sully" Sullenberger to remark, "What a view of the Hudson today!"[1]

Less than two minutes after takeoff, still rising and heavy with fuel, the plane hit a flock of Canadian geese. Sullenberger and his co-pilot, Jeffrey Skiles, saw them coming, but there was no time to avoid. "Birds," said Sullenberger. "Whoa," said Skiles, "oh, [explicative]." Even a single chicken will destroy an engine. In a moment, multiple large geese were sucked in, crippling the plane. Communication and other systems remained, but the thrust pushing the plane into the air was gone, leaving the crew piloting a fast-descending 150,000 pound glider. "Uh oh," said Skiles.

Sullenberger and Skiles had to take action quickly. Sullenberger contacted TRACON, the system of air traffic controllers that provides assistance in areas with multiple airports. "Hit birds, we lost thrust in both engines. We're turning back toward LaGuardia."

Air traffic controller Patrick Harten confirmed this decision and started preparing an emergency flight path and clearing runways. It's clear from the transcripts that loss of all engines was unprecedented. "He lost all engines," Harten said. "He lost the thrust in the engines. He is returning immediately." The reply from the airport tower was, with a tone of disbelief, "Which engines?!" "He lost thrust in *both* engines, he said," Harten responded. "Got it," the tower controller replied, and initiated emergency procedures.

It was only thirty-six seconds after announcing the bird strike that Captain Sullenberger gave up returning to LaGuardia. "We're unable. We may end up in the Hudson." The air traffic controller didn't immediately change the plan, and offered an open runway at LaGuardia. Sullenberger replied with one word: "Unable."

The air traffic controller acknowledged but offered yet another runway at LaGuardia. Sullenberger responded, "I'm not sure if we can make any runway. Oh, what's over to our right? Anything in New Jersey? Maybe Teterboro?" The air traffic controller contacted the Teterboro airport and confirmed an open runway to Sullenberger, all in less than sixteen seconds. During that time, though, Captain Sullenberger had assessed the situation and made his decision. "We can't do it," he said, exactly two minutes since the first emergency communication. "We're gonna be in the Hudson." He turned on the intercom to broadcast to the cabin.

"Brace for impact," he said.

Miracle or ...

Sullenberger's water landing and rescue of all 155 people on board has been called the "Miracle on the Hudson." But was it a miracle, or was it the product of decades of engineering and design choices, training regulations, and semi-autonomous systems incorporated into the brain of the plane itself? Certainly, luck played a role: The weather was clear. The accident occurred in the daytime. The Hudson River was nearby, fairly clear for a landing, and minutes from rescue boats. A cold front brought an unusual placidity to the Hudson, but without large chunks of ice in the water.

But it wasn't all luck. Many contributors to the Miracle were by design and under human control: Captain Sullenberger was highly experienced, with more than 20,000 hours of flight time since starting his commercial career in 1980. Before that, he was a fighter pilot in the US Air Force. He was also an expert in aviation safety, with his own consulting business and advanced degrees in Industrial Psychology from Purdue. First Officer Skiles also had over 20,000 hours of logged flight time and had flown at the rank of captain himself.[2] Although no runways ended up being used, Harden and the airport controllers communicated quickly and effectively. Everyone in the situation was consulting instrument panels, predictive systems, checklists, and displays throughout, each carefully designed to provide information and support fast decisions.

Thus, Sullenberger and Skiles landed safely because of their own judgments combined with the support systems engineered into the airplane and into the procedures they followed. Steven Johnson summarized the contributions of automation and decision aids poetically in his book *Future Perfect*:

> Most non-pilots think of modern planes as possessing two primary modes: "autopilot," during which the computers are effectively flying the plane, and "manual," during which humans are in charge. But fly-by-wire is a more subtle innovation. Sullenberger was in command of the aircraft as he steered it toward the Hudson, but the fly-by-wire system was silently working alongside him throughout, setting the boundaries or optimal targets for his actions. That extraordinary landing was a kind of duet between a single human being at the helm of the aircraft and the embedded knowledge of the thousands of human beings that had collaborated over the years to build the Airbus A320's fly-by-wire technology.[3]

However, an in-depth look into the Miracle also revealed problems with the flight systems. As we identify these problems, the landing on the Hudson becomes a perfect microcosm for how to make improvements in the human factors of future flights. The first order of business is to find the gaps between what pilots are capable of and what is demanded of them, and then what systems are or could be in place to close those gaps.

In the aftermath of the Hudson landing, Sullenberger and Skiles were scrutinized for their choices. Flight simulators were set up by the National Transportation and Safety Board (NTSB) and pilots tried to land on the Hudson or were asked to return to various airports. Some of these airport landings were successful, prompting gleeful headlines such as "Sully Could Have Made It Back to LaGuardia."[4] However, the simulator pilots knew what they would face ahead of time and reacted instantly, even then not always succeeding in landing at an airport. Only one attempt was made with a delay added after the birds were hit to simulate human decision time. When a delay was added, the pilot did not make it to an airport in the simulation – all onboard would have died.[5]

We tend to respond to critiques of our heroes by pushing back, or considering any questioning of them to be an insult. If the questions are meant to identify scapegoats and assign blame, then we are right to be upset when our heroes are questioned. Blaming "pilot error" on a scapegoat does little to prevent future incidents. But if we search for continuous improvement, then it is always correct to question those heroes to analyze what went wrong, what went right, and how we can make our systems even better. For example, the NTSB report noted that few pilots were able to hit the water at a good angle in the simulator, but that the one who did used a specific technique: "approaching the water at a high speed, leveling the airplane a few feet above the water with the help of the radar altimeter, and then bleeding off airspeed in ground effect until the airplane settled into the water." Thus, this new technique was learned from investigating the Hudson crash and is now taught to other pilots. Other findings by the NTSB were that there was no checklist for ditching a plane at low altitude – an issue the FAA then addressed. Trying to learn and improve is a core principle of good design, but one that is not as natural as the urge to blame. As an educated public, we must also insist that industries and governments adhere to the principle of learning and improving when creating products, regulations, or meting out punishments to the "bad apples."

"Caution! Terrain! Pull Up Pull Up!"

One of the devices designed to intervene in or before an air emergency is the Enhanced Ground Proximity Warning System (EGPWS). When crewmembers are distracted or have to attend to other issues, the EGPWS checks the aircraft position relative to the ground and obstacles. If the aircraft changes altitude too quickly, the system will verbally instruct "Don't sink!" If the aircraft is too close to the ground, particularly mountains, the system will start with information

"Too low, terrain," then instruct "Pull up, pull up" along with the reason why, "Terrain," and an alert word ("Caution!") to help the pilot know action must be taken. Terrain warning systems may be the biggest advancement in aviation safety since the 1970s – after their introduction, accidents involving "flying a perfectly good plane into the ground" dropped dramatically, almost to nonexistence.[6]

However, as with all human-engineered systems, the EGPWS is not always reliable. Most frequently it fails by offering false alarms in a loud distracting voice. If an airport is not programmed into the EGPWS (many are not) or the plane is making an emergency landing at a non-airport, then it will alarm constantly during the descent. Pilots are annoyed at the non-stop sound when landing at rural airports – and it is very very loud. The EGPWS can be heard in the cockpit audio from the Hudson landing because the system did not know the Hudson landing was intentional, adding more stress to an already stressful situation.

EGPWS: Too low, terrain. Too low, terrain. Too low, terrain. Caution, terrain. Caution, terrain. Too low, terrain. Too low, gear.
SKILES: Hundred and fifty knots. Got flaps two, you want more?
SULLENBERGER: No let's stay at two. Got any ideas?
HARTEN: Cactus fifteen twenty nine if you can uh . . . you got uh runway uh two nine available at Newark it'll be two o'clock and seven miles.
EGPWS: Caution, terrain. Caution, terrain.
SKILES: Actually, not.
EGPWS: Terrain terrain. Pull up. Pull up. Pull up. Pull up. Pull up. Pull up. Pull up. Pull up. Pull up. Pull up. (Repeats indefinitely in background)
SULLENBERGER: We're gonna brace.

As someone who has to turn down the radio to be able to concentrate when merging onto the highway, I felt for these pilots.

In a 2012 crash into Mount Salak, Indonesia, the crew had disabled the EGWPS system when they believed it was malfunctioning. All forty-five passengers died.[7] In a 2010 flight from Poland to Russia, with the president of Poland onboard, the Russian airport was not programmed into the EGWPS. The system sounded an alert as the plane approached, but it was ignored because the crew knew the airport was not programmed and they expected the alarms. Unfortunately, the alert was really about the trees and terrain they were going to hit before reaching the runway, rather than the "normal" alarms that they were landing at the airport. No one survived.[8] These accidents tragically illustrated the balance between human trust in the EGWPS automation and its reliability. When the systems are not trusted, they are turned off or ignored. At other times, the system may behave as designed, going off as altitude declines, but is a distraction from the emergency at hand. Finding the right balance depends on advance testing of these systems, because even the best automated system *will* fail. We can learn from these events for future designs, such as autonomous cars or drones. How failures occur, and how they affect the person in the vehicle or around it, is within our control. Will it fail gracefully, with back-up systems ameliorating the danger of the failure? Will it be transparent in its failure, so that the operator, pilot, or driver understands what is failing and when? Will it be obvious how to react to the failure, quickly and accurately? Answering these questions during design will give us the best chance at avoiding future tragedies.

Other decision aids in the cockpit included the electronics that partially automated flying the plane. The plane calculated and displayed the best gliding speed for Sullenberger and held itself to that speed, also displaying how it anticipated changing speed ten seconds in the future. This freed Sullenberger to focus on other decisions, rather than having to hold the plane to the right speed. Human factors psychologists and engineers call picking a person or machine for a job *function allocation*: gliding speed was allocated to the machine.

Function allocation means to consider what jobs best suit machines and what jobs best suit humans. For example, human reaction time is slow compared to computers, so when a fast action needs to be taken, especially if there is a reliable cue that prompts that action, it should be allocated to a machine. Harkening back to the EGWPS, checking an altitude boundary and issuing the command "Pull up" below that boundary is a simple job for a machine. Forcing a human to remember to check altitude while multitasking or in an emergency demands a great deal of time and attention. However, when it comes to visual pattern matching (e.g., picking out a target) or making decisions based on multiple ambiguous cues (e.g., diplomatic negotiation), humans have the edge. But making those kinds of decisions can be effortful, so it's often best to allocate as much of the lower-level decisions to a machine to free up the resources of the human for those tough questions.

Checklists Help with Decisions

Sullenberger and Skiles immediately went to the reignition checklist when their engines were destroyed and continued following it in tandem with water landing procedures, only stopping once the plane was in the Hudson. Following checklists is ingrained in aviation and is slowly becoming standard in other domains, such as health and medicine. However, even with the best checklist, humans may need to decide what and when to follow. When testifying in front of the National Transportation Safety Board, Sullenberger said,

> We didn't have time to consult all the written guidance, we didn't have time to complete the appropriate checklists. So Jeff Skiles and I had to work almost intuitively in a very close-knit fashion without having a chance to verbalize every decision, every part of the situation. By observing each others' actions and hearing our transmissions and our words to others, we were able to quickly be on the same page, know what needed to be done, and begin to do it.[9]

They also needed to prioritize their actions. In an interview with *Air and Space Magazine*, Sullenberger said,

> The higher priority procedure to follow was for the loss of both engines. The ditching [landing outside an airport] would have been far secondary to that. Not only did we not have time to go through a ditching checklist, we didn't have time to even finish the checklist for loss of thrust in both engines. That was a three-page checklist, and we didn't even have time to finish the first page. That's how time-compressed this was.[10]

Sullenberger's decision to ignore the three-page checklist, meant to be used at 30,000 feet instead of 3,000 feet, exemplifies the importance of usable design. In an interview with Jon Stewart on *The Daily Show*, the taciturn Sullenberger acknowledged how poor design reduced the information he could gather as the plane went down. Stewart commented with disbelief,

> You said your partner reaches over for the manual. I guess there's a manual they put in the cockpit for things that go wrong, and it used to have tabs on it for easy [access] – like, "blue tab is plane going down." And as a cost-saving measure they had removed the tabs. So he was literally going like, "I better check the index!" You know? Like, how crazy is that?

Sullenberger responded: "It's one of those minor things that by itself might not make a big difference. But, you know part of what we do is manage risk. We look for ways to make the system better. And I think that would make the system better, if we put the tabs back on."[11] Finding instruction quickly (in Sullenberger's case more quickly than anyone had imagined) could have been easier with search tabs, or a fast electronic search, or a just-in-time display fed by artificial intelligence. The ways to support are as unlimited as human imagination paired with engineering – but all of them should be tested for ease of use, especially in time-critical emergencies with other alarms and systems going off.

Checklists Put Everyone on the Same Page

Checklists ensure an entire team understands the past, present, and anticipated future of their job. This shared mental state requires *theory of mind*, a term first coined by developmental psychologists to describe how small children move from having their own thoughts to understanding that other people *also* have thoughts, and those thoughts can be different from theirs. It is hard to imagine now, but when we were small children we did not know that the people around us could be thinking, knowing, or seeing things differently than we do. Incidentally, this also means small children are terrible (even incapable) liars, since they believe you already know everything they are thinking. It also explains their frustration when you don't seem to be able to read their minds to know what it is that they want.[12]

Theory of mind is not just for kids. It persists in adulthood in small ways. Adults understand other people have their own thoughts and experiences, but we still tend to believe others think more like we do than is true and are often shocked when confronted with just how differently another person thinks. Just read the comments section on any news site. Research studies on adult theory of mind often have a person try to communicate an idea to another (one did this through drumming out a song on a table) and have them judge how well they believed the other person understood their intent. Then, the researchers compare that judgement to the other person's actual understanding. In the drumming study, people overwhelmingly thought the other person would "get" the song – it seemed so obvious to the drummer, how could it not be obvious to the receiver? But it was not obvious. Hardly any receiver guessed the correct song from hearing the drumming (the exception being "Jingle Bells").[13] This one you can try at home.

Checklists in medicine can help prevent theory of mind mistakes from impacting surgical outcomes. I worked with a team of veterinary cardiologists in 2015 to develop a surgical checklist for their procedures and one important step in the checklist was for the surgeon to review "anticipated critical events and unexpected steps" and "expected operative duration" with the rest of the team. This was because the surgeon fully understood the medical history of the animal, including age and other potentially complicating variables like body size or severity of the heart problem. It would be easy for the surgeon to think the whole team shared in all of this knowledge and expectation, but there was a wide variety of knowledge, experience, and roles distributed among the surgical technicians, anesthesiologist, and anesthesia technicians. An explicit call for communication in the checklist made sure everyone was working with the same information, had similar expectations, and could make better decisions based on that information.[14]

Thus, human decisions are aided in two ways by a checklist. Checklists support fallible human memory and encourage explicit communication within a team. The quote from Dr. Atul Gawande that began this chapter captured the messages of our failures, as one step or job may be simple enough, but complex systems such as aviation and healthcare overwhelm the human brain.[15] It is the formal systems we create, such as checklists, that address the "profound difficulties" of making them work.

Conclusion

The secret to the survival of our species is our adaptability. We have created an impossible technological world, one where we shouldn't be able to function, where machines carry us too high for oxygen in the atmosphere and too fast for our reaction times. Yet, we have also created systems to support our capabilities and overcome our limitations. This means that our heroes are not superheroes, nor do they need to be. They are human. The intense training for pilots, surgeons, and police is important, but it is augmented by the technology that supports their decisions and actions. Acknowledging that we're not perfect, that we need support from checklists or automated systems, is the first step to being able to accomplish the extraordinary, such as keeping millions of flights safe in the air each day.

The next step is to make sure those support systems and automations are designed to fit with our all-too-human limits: checklists can't be too long, automation needs to account for failures and false alarms, and we need repeated reminders that other people think differently than we do. We need to keep an eye on the technology – if it's poorly designed, it can do more harm than good. We must insist on, and dedicate resources to, a culture of improvement, where we analyze what goes wrong but also what goes right. The Miracle on the Hudson illustrated every aspect of such a culture, from the well-tested and helpful fly-by-wire systems on the plane, to the after-incident investigation and tests, which improved training and interfaces for future flights.

2 BAD WATER

> Nutshell: Agitators claiming "environmental racism" and comparing Flint water to the gas chambers at concentration camps. But Flint water still determined to be safe.
>
> – Davis Murray, March 3, 2015, Michigan Governor's Office

How to Believe You're Right, Even When You're Wrong

What do astrological signs, the Iraq war, and politicians in Flint, Michigan have in common? The people involved looked for evidence that confirmed what they already thought they knew: Virgos are hard workers, Saddam Hussein had weapons of mass destruction (WMD), and the water looks nasty, but it's safe to drink. Looking for support of what we want to believe, or what we already believe, is so natural we don't even know we are doing it. It's so natural our memories are hard-wired to back us up when we do it – it's easier to notice and remember evidence that fits with what you already believe. Maybe the last time you hired a hard worker you asked if they were a Virgo, and when they said yes, you delightedly exclaimed, "I knew it!" Your brain keeps the tally of those successes. But whenever you were wrong, you silently forget that inconvenient fact. We "remember the hits and forget the misses." We are biased toward *confirming* our previous beliefs and remembering all the times we think it was right to do so.

Virtually every web search is affected by confirmation bias. If you search for the name of a politician followed by any adjective (courageous, stupid, genius, original, blowhard, eloquent) you will find a website that declares a candidate has that attribute. The same is true of health questions – search for any symptom and it's guaranteed you'll find it comes with the possibility of cancer. To check the truth of this myself, I promptly found a webpage called "Can a stubbed toe lead to cancer?" Hypochondriacs do not stand a chance with the Internet at their fingertips. With a world so full of information, knowing what sources to trust becomes a harder question than finding the information itself.

Whether someone is a Virgo hardly matters, but unfortunately there are some serious examples of confirmation bias changing the world for the worse. (Of course, I could be guilty of confirmation bias right now: I only notice when the world changes for the worse.) When the residents of Flint, Michigan learned in 2015 that their water had been poisoned with lead, there were many factors at play. The events leading up to the public announcement and the government response was a tragic case study of confirmation bias.

Confirmation Bias as Poison

> The act, and the program here in Michigan, work to ensure that water is *safe* to drink. The act *does not* regulate aesthetic values of water.
>
> – Email to Gov. Rick Snyder from his Press Secretary, February 1, 2105[1]

Some Flint residents had brown water coming through their taps. Others had yellow. It smelled bad. In late summer of 2014, residents had been through two "boil water" advisories in less than a month. After each one they were told that *now* the water was safe to drink. It might not look pretty, but it was safe to drink. We know now that it wasn't – and the problem wasn't with bacteria that could be killed with heat, it was lead contamination.

Although many variables contributed to poisoning Flint residents with lead, it took so long to address the problem because government officials believed there was no problem. According to government response and internal emails, they only considered reports and data that confirmed that belief while ignoring those that disconfirmed it.[2]

The competing sources of information were: the Michigan Department of Health and Human Services (MDHHS), the Michigan Department of Environmental Quality (MDEQ), the Environmental

Protection Agency (EPA), Mona Hanna-Attisha (a pediatrician representing a group of Flint pediatricians), Marc Edwards (a civil engineering professor at Virginia Tech), and the people of Flint.

First, a refresher on the engineering and scientific facts of the tragedy. The water pipes in Flint, as in many US cities and towns, are made of lead. These pipes are safe, as long as there is a protective layer between the water and the lead. Lead levels increased in Flint because the city moved from using Detroit water to using river water. River water was cheaper, but also more corrosive. It ate away the protective layer, allowing the drinking water to leach lead from the pipes. Putting a safe-to-drink chemical in the water would have prevented the corrosion,[3] but the MDEQ approved omitting it. They blamed a misreading of the "Lead and copper rule," federal guidelines from the EPA, when their decision was uncovered.[4]

The Michigan agencies vigorously tested the water after the switch, but their testing minimized the lead levels by flushing the pipes before testing and sampling at a low flow rate.[5] An EPA official (Miguel del Toral) voiced early concerns purely based on his knowledge of the process Flint was using. He knew the corrosive river water needed to be treated as it flowed through lead pipes and that this was not being done. He also noted that testing agencies should not flush before testing, and that *any* lead in the water after flushing was a sign that much more was present. Last, the city had added another chemical (ferric chloride) that made the already corrosive water even more likely to leach lead.[6]

The Michigan officials also missed the high levels because they looked at the data for the whole area, rather than by postcode. The overall levels didn't look too high until you divided by postcode and found that just a few postcodes were quite high. The Flint pediatrician Mona Hanna-Attisha and colleagues found the high lead levels by postcode and tried to bring this to the attention of the officials, but to no avail.[7]

Outside help came at the call of a citizen-scientist, LeeAnne Walters, who tested her own water to understand why her children had rashes. She then contacted Marc Edwards, an expert in lead testing. When Edwards and his team tested water in homes, they found lead levels high enough to classify the water as "toxic waste." They were shocked and repeated the tests, but found identical results. These findings were combined with the high lead levels in children detected by Hanna-Attisha to make a strong case for lead leaching into the water of Flint, especially in certain geographic areas.[8]

But the MDHHS already "knew" that the water was safe. They found data that showed spikes in children's lead levels every summer, with no difference in levels after the switch to corrosive river water. They directly contradicted the pediatrician's findings: "[The pediatrician group] used two partial years of data, MDHHS looked at five comprehensive years and saw no increase outside the normal seasonal increases."[9] The Michigan DHHS concluded there was nothing to worry about, especially paired with the low lead levels measured by the MDEQ (taken after flushing and possibly from a mixture of lead and non-lead pipelines). But with the warnings from del Toral and the evidence from the pediatricians and Edwards, why did officials only consider data that told them lead was not a problem?

Released emails give a picture of how state officials viewed the town of Flint: a poor locale with poor people complaining about perfectly safe drinking water. Officials said the residents were overly concerned with the "aesthetics" of the water (color, taste) instead of safety.[10] Aesthetics did seem to be at the forefront, as residents had started bringing bottles of brown and orange water to wave in protest at their elected officials. From a *New Yorker* article, emails from officials said, "The 'aesthetics' ... were bad because 'it's the Flint River'; 'the system is old'; 'Flint is old' – the water, in a word, fit their picture of the city."[11]

No one wanted to maim or kill the children of Flint. But believing there was no problem was easy to support. Officials only looked at data they trusted from local and state sources, while discounting information from federal agencies and academics they assumed had their own agendas. An email from the chief of staff to the governor said as much: "Now we have the anti everything group turning to the lead content which is a concern for everyone, but DEQ and DHHS and EPA can't find evidence of a major change ... Of course, some of the Flint people respond by looking for someone to blame instead of working to reduce anxiety."[12] He also wrote, "The DEQ and DCH feel that some in Flint are taking the very sensitive issue of children's exposure to lead and trying to turn it into a political football claiming the departments are underestimating the impacts on the populations and particularly trying to shift responsibility to the state."[13] An email from the head of the MDEQ to a reporter said of Edwards, who had also found high lead levels years earlier in DC,

> this group specializes in looking for high lead problems. They pull that rabbit out of that hat everywhere they go. Nobody should be surprised when the rabbit comes out of the hat, even if they can't figure out how it is done ... [W]hile the state appreciates academic participation in this discussion, offering broad, dire public health advice based on some quick testing could be seen as fanning political flames irresponsibly.[14]

Because the MDEQ had already made their decision, it was easy to assign motives and reasons to information that conflicted with the decision. Ironically, officials blamed Edwards of the same cognitive bias they were suffering from, saying he expected to find lead wherever he looked, and that's why his tests found it.

Complex decisions often involve ambiguous data. It is easy to only look for data that support what we already believe. We also get too invested in our prior beliefs and it is difficult and sometimes painful to change them. The only way to prevent confirmation bias is to systematically search for, include, and consider data *against* our beliefs. This can be done with a decision aid, where evidence for and against a position is required prior to the final decision. It's good practice for everyone to do this, but for important policy-level decisions it should be mandatory.

Half-Empty or Half-Full: It Still Has Lead in It

> Political language ... is designed to make lies sound truthful and murder respectable, and to give an appearance of solidity to pure wind.
>
> – George Orwell, 1946, *Politics and the English Language*

Ambiguous data may be the first catalyst for bias, but it's often followed by another strong influence: the framing of the question. Should the government take four million dollars worth of action if there was an 80 percent chance the Flint water was safe to drink? Should they take action if there was a 20 percent chance the Flint water could permanently harm the children of Flint? The obvious answer to both questions is "yes" and the math is the same, but each could lead to different answers from a government looking to confirm evidence the water did not contain lead.

The most famous and often-repeated study on framing evidence was designed by Amos Tversky and Daniel Kahnaman.[15] They had people choose one of two options for a government health initiative. They were told to imagine that the USA is preparing for the outbreak of an unusual fatal disease, which is expected to kill 600 people. Two alternative programs to combat the disease have been proposed. Assume the exact scientific estimate of the consequences of the programs are as follows:

The first group chose between:

- Program A: "200 people will be saved"
- Program B: "There is a 1/3 probability that 600 people will be saved, and a 2/3 probability that no people will be saved"

The second group chose between:

- Program C: "400 people will die"
- Program D: "There is a 1/3 probability that nobody will die, and a 2/3 probability that 600 people will die"

When you look at all four groups, it's obvious that A and C are equal, math-wise. They both guarantee an outcome. Programs B and D are also equal – they gamble with a potential 100 percent win and a potential 100 percent loss. So, we'd expect that the same percentage of people in each group would choose A and C compared to B and D.

This is not the case. The first group is presented with a positively "framed" situation. The programs talk about saving people. In this case, about 72 percent of people choose A over B. In choosing the guaranteed outcome, we say that those choosing A are more risk-averse. We want to save people, not take risks.

But in the second group, only 22 *percent* of people choose program C, despite it being identical to program A! This was because the programs for the second group were framed as people dying rather than being saved. No one wants to choose that people die. This negative frame put people in a more risk-taking mood, and then they tended to choose the gamble of program D over C. The short take-home message is twofold: (1) phrasing changes people's decisions and (2) positive phrasing makes people risk-averse while negative phrasing makes them more likely to take a gamble.

Advertisers and political lobbyists are certainly aware of this research:

- We hear "Four out of five dentists agree, use Brand X" not "One out of five dentists says to avoid Brand X"

- Which makes you more angry? Paying for a $10 parking pass or getting a $10 parking ticket?
- This was an actual ballot measure in 2014: "Shall the Missouri Constitution be amended to ensure that the right of Missouri citizens to engage in agricultural production and ranching practices shall not be infringed?" Consider a different framing (that is just as accurate): "Shall the Missouri Constitution remain unchanged to apply long-standing rules concerning the environment and animal treatment protections?" (The amendment was passed after a recount.)[16]
- If you want people to buy insurance, you might stress that "Not buying insurance will put you at risk for losing everything after a medical emergency" rather than "Buying insurance will keep you solvent after a medical emergency."
- People rated the quality of meat described as either "75% lean" or "25% fat" – ratings were higher for the 75 percent lean.[17]

Do framing effects translate to real life beyond buying meat or toothpaste? Yes. When patients were told a vaccine was "70 percent effective" they were more likely to accept it than if they were told it was "30 percent ineffective."[18]

In a study that should terrify the American public, Professor Valerie Reyna and her colleagues found that US Intelligence Agents were even more likely to be biased by the way a question was framed compared to college students or a group of adult citizens. They were also more confident that their decisions were the right ones. My favorite line from the article was, "During the course of this study, these agents expressed motivation to demonstrate their professional expertise in decision making."[19] Clearly, motivation isn't enough when framing effects are at play.

What can we do? First, we must recognize that language and phrasing are important. When I used the "Flint water crisis" example of confirmation bias in a graduate class, one astute student pointed out that this label framed the problem as one without victims or perpetrators. A crisis for whom, she asked? Brought about by whose decisions? As the "Flint water crisis" it merely exists, like a tsunami or hurricane. A better phrasing would be, "Governmental mistakes caused lead poisoning of the people of Flint." This identifies cause and victims. Of course, in the beginning of a crisis, until direct causes are known, one cannot describe it with such certainty. It's understandable that the media desires a shorthand label like "water crisis." However, that label can unintentionally reframe causes and de-emphasize suffering. As we learn more, as we have in the case of Flint, we should rephrase our labels to accurately reflect the evidence. And being aware of the power framing can have, we can push back and insist on accurate labels.

The second action we can take to reduce framing effects is to require justification for decisions. Although everyone succumbs to framing effects, one study found that older people are even more susceptible.[20] One intervention used in research is to make people justify their decisions. This reduced framing effects across the board. After justifying their decisions (and often revising them during the justification), the older and younger people in the study were all equal. We should require the same justification from anyone making decisions that affect the public, whether it's recommending a new flu vaccine or changing a decades-old water system for an entire town.

Known Knowns

The CIA began employing rules about seeking disconfirming evidence after no WMD were found in Iraq during the 2003 US invasion. Early in the war, the US Secretary of Defense, Donald Rumsfeld, famously said, "There are known knowns. There are things we know that we know. There are known unknowns. That is to say, there are things that we now know we don't know. But there are also unknown unknowns. There are things we do not know we don't know."[21] The unknown unknowns were clearly made out to be the bad guy in this list. But what the USA was facing at the time was more of a problem of mistaken knowns, those "known knowns" that turned out to be wrong, especially the assertion that Iraq had nuclear or chemical weapons prepared to launch on a large scale.

The US Department of Defense would spend the next decade analyzing how so many people could come to such a wrong conclusion. Even in late 2003, a CIA press release was still saying that WMD would be hard to find, so just because we hadn't found any didn't mean there weren't any.[22] Looking back many years later, nothing was ever found. As Ambassador James Dobbins said in an interview with *Foreign Policy* magazine in 2013, "The problem was that, with the Iraqi WMD, the policymakers wanted bad news. They wanted to confirm that Iraq had WMD, and the intelligence analysts were inclined to move in that direction anyway, since it would be even worse if they predicted they didn't have WMD and it turned out they did."[23]

This particular confirmation bias-based intelligence failure made it clear we needed better training for intelligence analysts. They needed help in turning large amounts of data into reliable information and how to avoid our natural human biases. After all, as mentioned earlier in the chapter, these analysts are as susceptible to bias as everyone else (or more susceptible). The Intelligence Advanced Research Projects Activity (IARPA) organization was founded in 2007 to find a solution. Cognitive biases such as confirmation bias are notoriously difficult to train away, which is why procedures needed to be employed to mitigate their effects. Some of these mitigation procedures are high tech, like computer games giving personalized feedback on biased decisions,[24] but one of the most successful can be done on paper: an Analysis of Competing Hypotheses (ACH), developed by CIA analyst Richard Heuer.[25]

How to Stop Being Biased

In an ACH, an analyst follows steps to formalize decision-making. The first step is to write down all possible hypotheses, even ones that seem unlikely or directly contradict what the analyst thinks they know. For critical analyses, a diverse team of analysts should work together on this in a brainstorming session. Second, all evidence is written down next to the hypotheses it supports and the hypotheses it contradicts. Third, disproving a hypothesis is more valuable than supporting a hypothesis, so the analyst should take each piece of evidence and see how many hypotheses it eliminates. Elimination is the goal, not supportive evidence.

With the reduced amount of hypotheses available, the analyst notes what information would be needed to disprove the rest of the hypotheses and goes searching for it. This step is incredibly valuable because it works in direct opposition to confirmation bias. It forces the analyst to look for evidence *against* every hypothesis instead of for. Tentative conclusions can be reached at this stage once the most unlikely hypotheses are rejected. Then a "what if" analysis is performed that states how the current conclusion would be changed if any of the evidence presented (the known knowns) were wrong. Last, the analyst derives conclusions but must also present the other hypotheses and why they were rejected.

Here is an example of an ACH I did to analyze where the puddle on my kitchen floor came from (see Table 2.1). Where did it

Table 2.1 Who's guilty? A de-biased investigation

	Hypotheses					
Evidence	**My spouse dripped water while cooking**	**My spouse spilled his drink**	**The dog pottied in the house**	**I spilled my drink**	**It's a leak**	**The dog threw up**
Dinner was an hour ago	VC	C	N/A	C	N/A	N/A
We had wine with dinner	N/A	C	N/A	C	N/A	N/A
Liquid smells bad	I	VI	VC	VI	C	VC
Liquid is almost clear, slightly yellow	I	VI	VC	VI	C	C
Puddle appears about 4 oz	C	C	VC	C	C	C
Puddle located near door	VI	C	VC	C	VI	C
Puddle is watery, not viscous	VC	VC	VC	VC	C	I
Puddle is greater than 4 ft. from any appliance	I	C	N/A	C	VI	N/A
Dog has not requested to go outside in "a while"	N/A	N/A	C	N/A	N/A	N/A
Dinner was chicken wings and sautéed greens (no boiling or steaming)	VI	N/A	N/A	N/A	N/A	N/A
Score	7	4	0	4	4	1

VC = very consistent, C = consistent, I = inconsistent, VI = very inconsistent come from? Who gets the blame? Let's look at an ACH to make our best guess.

First, I generated multiple hypotheses, based on my knowledge of the situation. Perhaps it came from cooking. Or maybe the roof is leaking. Then I started to collect evidence to eliminate some of those hypotheses. Just like in 5th grade math, it's important to "show your work" here. Some of the evidence was based on time ("Dinner was an hour ago"), some on location ("Puddle was near the door"), and some on the attributes of the puddle ("Liquid smells bad").

All the evidence here is credible, but if we had some evidence we weren't sure about, we could give it more or less weight. Each hypothesis is rated from Very Inconsistent (VI) to Very Consistent (VC) with each piece of evidence.

Looking at Table 2.1, we want to see which hypotheses we can refute based on the evidence. "Very inconsistent" receives two points and "Inconsistent" receives one point, so we are looking for the highest scoring hypotheses to eliminate. It looks like the chef is in the clear with a high of seven points, and us spilling our drinks or an appliance leak is also unlikely. However, it's difficult to determine the exact nature of the puddle and I would present both hypotheses (potty dog and sick dog) with the supporting evidence. Either way, we now have a suspect. Based on the ACH, the Brittany spaniel was summarily judged and sentenced to the usual punishment in my house: a stern look with a sigh followed by treats and a good belly rub.

Imagine if those in charge of deciding what to tell the people of Flint had used the ACH. They could include their hypothesis that the river water was safe, if unattractive and smelly. But they would have been forced to list other hypotheses: The water has harmful bacteria in it. The water has harmful chemicals in it. The water has harmful minerals in it. The testing done by Edwards was accurate. The postcode-level data gathered by the pediatricians revealed lead was only worse in some areas. Remember, these should be developed by a diverse group – not just one or two Michigan government officials. Then they would have to collect the evidence that helps them reject each of those hypotheses. I will grant that oversight of the research methods was also important. If they chose methods that were almost guaranteed to produce results with no lead, such as flushing the pipes before measuring levels, the ACH could be misused. But overall, it forces two important processes: Consider explanations outside one's prior beliefs and search for evidence to disprove, rather than prove, a hypothesis.

There is no method that guarantees successful prediction or analysis, but ACH appears to minimize many of our cognitive biases and provide an easy-to-reference roadmap for how the decision was made. The main drawback is the time required; this is not a method for hair-trigger decisions. However, for longer-term decisions, development of tools like this is required to override our inner self, who has the best of intentions, but rarely makes true evidence-based decisions.

Conclusion

We get by. Except when we don't.

Our decisions range from good to terrible, usually without major consequences. When we do make bad decisions, it's usually in a predictable way, following well-known biases. We can deploy decision aids to help us manage even with limited experience and limited memory, but have to fight our inclination that we don't need such aids.

Policymakers, government officials, and intelligence analysts make life-or-death decisions, but they are subject to the same biases as normal people. We need them to take advantage of all of the research in human decision-making to make the best choices for us.

3 HYBRID VIGOR

> Increased vigor or other superior qualities arising from the crossbreeding of genetically different plants or animals. Also called heterosis.
>
> – *American Heritage Dictionary*, 2020

> Creativity emerges out of difference, in particular differences between people. It is because individuals have unique experiences of the world, particular life trajectories, and develop personal sets of knowledge and skills, that their perspective on a creative task is likely to be dissimilar, at least to some extent. Sharing different perspectives and reflecting on the difference between them plays an important part in the creative process.
>
> – Vlad Glăveanu and Marie Taillard, 2018[1]

Rate your creativity for each activity described in these statements from 1 to 5, 1 meaning you're much less creative than the norm and 5 meaning you're much more creative than the norm:

Appreciating a beautiful painting.
Finding something fun to do when I have no money.
Planning a trip or event with friends that meets everyone's needs.
Taking apart machines and figuring out how they work.
Making a scrapbook page out of my photographs.

These statements are part of a creativity test designed by James Kaufman, a professor at California State University. The higher the sum of your points, the more creative you are.[2] Creativity isn't even limited to the person; groups can be creative too, as they brainstorm new designs and solutions to problems. "Creative" is universally a compliment and we tend to label people as being creative or not, as though it is some permanent state. However, a great deal of research shows that there are techniques to encourage creativity and that we can recognize when we are trapped by fixed thinking (and escape!).

Groundhog Day

Deciding to solve a problem is one thing, but deciding *how* to solve it is another. We do the best when a new problem is almost identical to a previous problem because our bias is to try and solve new problems like old problems. Psychologists call this a *set effect*, meaning that you're stuck in one set way of thinking and can't move on to another, just like the film *Groundhog Day*. Also, the more often you're successful with a strategy, the more doggedly you stick to it – that set becomes stronger and stronger.[3] In 2011, I toured the 1663 Bronck House in Coxsackie, New York. The tour guide led us through incredibly preserved architecture – especially considering the Bronck family lived in that same house until the 1930s. The home has parapet walls higher than the roof connecting them, called a Dutch gable. This way of building looks nice, but it's difficult to prevent leaks at the point where the roof hits the walls, especially before modern-day flashing and waterproofing. The guide at the Bronck House vividly described how the home had water running down the walls, and that the inhabitants would stuff rags or other detritus into the gaps between the walls and roof. In this particular house, they did this literally for hundreds of years. An overhanging roof would prevent this problem, but even in the rainiest climates there are old Dutch homes that used this design (Figure 3.1). Apparently, the Dutch weren't going to change how they built houses just because of a little water running down the walls![4]

I came across a modern example of overcoming set effects when a friend told me about some creative problem-solving at his workplace. He ran a data center that needed constant cooling, but the power went out. This triggered a backup diesel generator to start up and keep the air

Figure 3.1 The Bronck House, occupied from 1663 to 1938. Hudson Valley, NY. Note how the side walls extend higher than the roofline.

conditioning running. Unfortunately, once it started up no one could get it to turn off. In his words,

> After a power outage, the folks at the data center were trying to get the diesel generator, which was as large as a pickup truck, to shut off once the utility power came back on. It wouldn't shut off, because the control unit was broken. So, they tried to force it to shut off by triggering a safety cut-off. If the air intake was blocked, it would automatically shut down. They tried a bunch of things, including stuffing a towel over the air intake to make it stop, but they could not get a good enough seal to cut off its air supply and shut it down. So, one guy gets the brilliant idea of taking his shirt off and smushing his stomach (which was a little paunchy) over the air intake. Presto, the seal was good enough and the generator stopped so they could start trying to repair it.

Using a stomach as an air-intake plug requires one to consider the attributes of the stomach apart from the normal purposes of a stomach. (Presumably these are to hold food, to keep one's shoulders away from one's pelvis, and occasionally serve as a cup rest.) When people can't think outside the usual context for an item they are *functionally fixated*; this is a particular type of set effect. When we are fixated on the typical function of an object, we cannot think of novel uses. Using a large wrench as a hammer is easy – a wrench shares so many attributes with a hammer it's natural to hit something with it. But using a can of vegetables as a hammer? That requires you to refocus on what makes a "hammer" (heavy, held in grip in the hand, can swing, flat striking plate surface a good distance from the hand). Modern jargon calls such solutions a "hack," but the term is anything but derogatory. My favorite story of such a hack came from a friend who, on a road trip in a car with non-working windshield wipers, was facing driving through a thunderstorm. She and the passenger tied bikini tops to each wiper and alternately pulled them back and forth through the almost-closed side windows to clear the glass.

The most famous hack of all time is probably Alexander the Great "untying" the Gordian knot by slicing it in two with a sword. But only slightly less famous is the genius engineering of the ground crew during the Apollo 13 crisis en route to the moon, where a carbon dioxide scrubbing machine was made to work using only materials the astronauts happened to have with them.

Square Pegs and Round Holes

Just hours into the Apollo 13 mission, an oxygen tank exploded, destroying access to electricity, water, and light. The astronauts retreated into the tiny space of the lunar module, an independent vessel inside the main craft. The lunar module was built for only two people to use for thirty-six hours, but suddenly it was a life raft supporting three astronauts for the over ninety hours needed to get them back to Earth.

In addition to the cold (38°F/3°C), the wet (condensation dripping down the walls and in the interior of the electronics), and lack of water (they ate all their foods containing water and drank only 6 oz a day to conserve it as a coolant for the ship), the three astronauts were facing certain death by carbon monoxide poisoning. The lunar module scrubbers were full and no longer taking carbon monoxide out of the air, but the only other scrubbers were made for an entirely different machine and would not attach inside the lunar module.

On the ground, engineers worked with the materials they knew were available to the astronauts, trying to build an adapter:

- Cardstock cover to the Apollo 13 flight plan
- Roll of gray duct tape
- Two liquid-cooled garment bags
- Two hoses from suits
- A sock (they ended up using a towel)

These materials were cobbled together into a filter that incorporated the square, differently sized scrubbers into the round air scrubbing machine in the lunar module (Figure 3.2). Almost every item was used for an

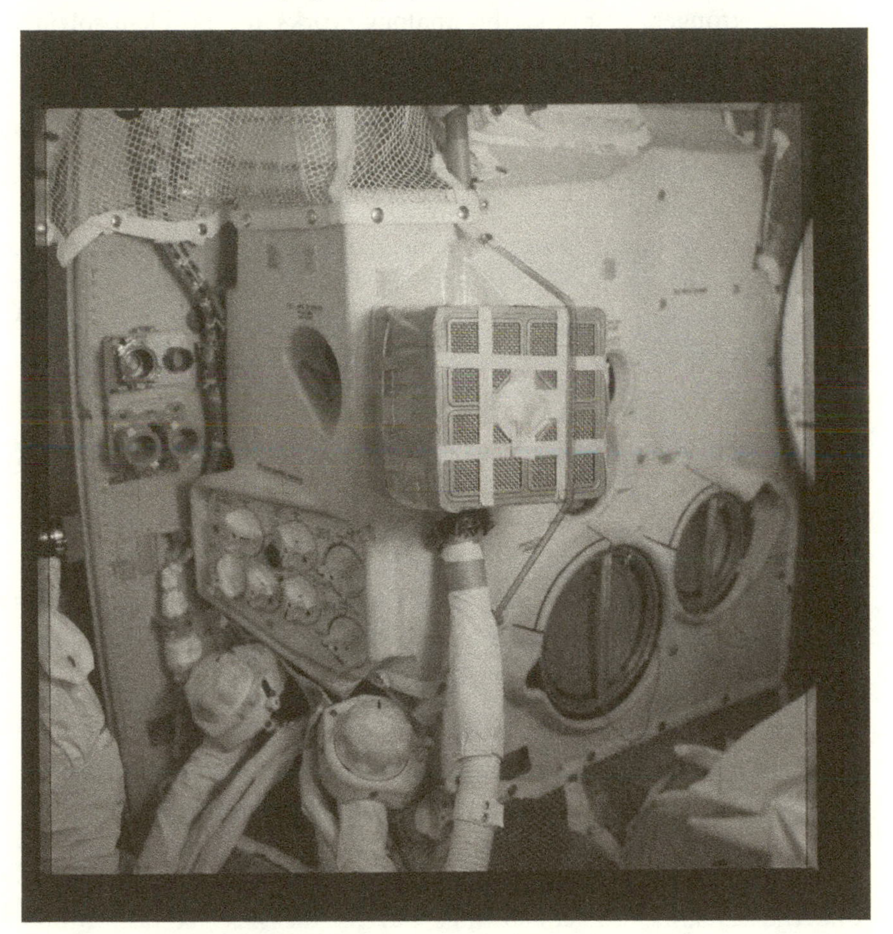

Figure 3.2 Literal outside-of-the-box thinking that saved the astronauts of Apollo 13. The box and hose coming from the bottom form the adapter that allowed use of the misfit carbon monoxide scrubber.

unforeseen purpose, particularly the flight plan cover page and the bags their special liquid-cooled clothes came in!

The scrubber worked. Returning to Earth still meant many hurdles, but at least they could breathe. Apollo 13 landed safely in the Pacific Ocean on April 17, 1970 and was dubbed "a successful failure."[5] Creative problem-solving saved three lives.[6]

Hybrid Vigor in Thinking

Summing up set effects links back to the title of this chapter: hybrid vigor. Although strictly meaning that diversity in a gene pool can produce stronger offspring, the analogy works for problem-solving. Bringing in diverse thoughts, broadening thinking about the functions of items, combining old tools together in new ways, all increase creativity and reduce the chance of being stuck on a problem or in a set way of thinking. The philosopher John Stuart Mill noticed this long before any scientist took it on as a research project. In his 1848 book on economics and the value of foreign trade, he lauded the financial value of diversity, saying, "It is hardly possible to overrate the value, in the present low state of human improvement, of placing human beings in contact with persons dissimilar to themselves, and with modes of thought and action unlike those with which they are familiar," and:

> Such communication has always been, and is peculiarly in the present age, one of the primary sources of progress. To human beings, who, as hitherto educated, can scarcely cultivate even a good quality without running it into a fault, it is indispensable to be perpetually comparing their own notions and customs with the experience and example of persons in different circumstances from themselves …

Mill, an ardent anti-slavery advocate, was always willing to throw shade on his peers during the nineteenth-century "low state of human improvement."[7] As my students would say, "He's not wrong."

Hybrid vigor also applies to combining our reasoning abilities with our emotions. It's hard to recognize how much emotion is affecting our choices and problem-solving, but as the next section shows, research has helped to outline when emotion drives our choices, even when we think we're acting logically.

Spock versus McCoy

Would it be better if everyone's choices were based on evidence and logic? What would the world be like if emotion weren't a factor in important decisions? In the original *Star Trek* series, the emotional Dr. McCoy was pitted against the impassive Mr. Spock, each one believing his approach to be the correct one. Our brains fight similar battles, weighing logic and emotion to decide where to live, what to buy, and who to be.

It's complicated to consider whether logic or emotion is "best" for a decision because most decisions we make have ambiguous outcomes. I was happy with the last apartment I chose to live in. But was it the best choice? Was there a much better apartment out there? I'm afraid to fly, but willingly get in my car to go to a grocery store, knowing logically that I'm much more likely to die in a car wreck. If you work for a company are you and your co-workers the best team for the job? It's hard to judge if we made the best decision. But even if we don't know what the best decision would have been in these situations, we know that many decisions are not optimal. Driving from Raleigh, North Carolina to Washington, DC rather than flying because of fear of a plane crash is neither optimal nor rational. The most important question for researchers is whether it's possible to predict these suboptimal decisions. Indeed, people make a lot of wrong decisions, but not at random. We are affected by both the physical signals sent by our bodies to our brains and by predictable biases in problem-solving.

Our Bodies, Ourselves

Philosophers often discuss the independence of the mind from the body. Much of philosophy is not open to experimentation, but this question is. Spoiler alert: the answer will not sit well with notions of free will and independent thought. Our seemingly conscious decisions are heavily influenced by the physical signals from our body.

A few examples: Judges are more likely to give a longer sentence for the same crime before lunch than for cases after lunch.[8] Doctors are more likely to prescribe opioids to patients at the end of the day compared to the beginning.[9] When people have their adrenaline amped by a challenging video game, they are more likely to think their partner in the game deserves to receive a shock for conflicting with them.[10] In all of these, the decision-maker is influenced by the signals their body sends (hunger, fatigue, excitement) instead of just logic and facts. None of these people knows that their decisions are being affected by their physical bodies.

In high-pressure fast-changing situations, this emotional decision-making can be necessary. Approaching the edge of a cliff creates feelings that make one want to back away. Emotions cut down the possibilities for action, making it a decision between a few options instead of hundreds. Psychologist Antonio Damasio developed a hypothesis for why we need emotions for decisions and called it the *somatic-marker hypothesis*.[11] The idea is that we receive signals (markers) from our bodies (soma) and they focus our choices. That funny feeling you get in the pit of your stomach is information being sent to your brain, not the other way around.

In modern life there aren't many life-threatening events day to day. I can't remember the last time I had to run from a baboon or battle my neighbor for dibs on the buffalo carcass. Our brains are no different than they were 30,000 years ago, but now most emotions aren't related to physical safety. Consider how parents feel when their child's behavior is criticized by an outsider, or the impact of a co-worker's judgmental tone or mood. No one is going to be eaten by their boss, no matter how scary the boss's reputation, but skyrocketing emotions in modern social situations have the same effect as life-or-death threats on our ancestors. Emotions suppress logic and push us to lash out disproportionately. For those living with disorders such as borderline personality disorder, over-the-top emotional reactions to perceived threats can harm their relationships and careers. Author Stacy Pershall, in her fantastic memoir *Loud in the house of myself*, described living with these extreme feelings: "You give your feelings melodramatic names, grandiose status, because melodramatic and grandiose are how you're feeling. You're the most depressed person EVER, or on the rare good days, the happiest – no, not happy, ECSTATIC. There is no gray, there is only the blackest black and shimmering white."[12] It is no wonder that we seek to tamp down the extremity of these emotions to better function.

Researchers and therapists have developed strategies to help people train themselves to better understand and control their emotional reactions. These therapies teach people to acknowledge their emotions are rising and that they feel threatened, but to let those emotions die down before taking action. This includes forcing a pause before reacting, removing oneself physically from the situation for a bit, and accepting an emotion but not acting on it.

Suppressing the overflow of emotions is a skill that takes practice, but once gained it can stop the explosions and regret that come from overreacting. Those suffering from emotional disorders are just experiencing an extreme version of the reactions most people feel due to everyday stresses. In recent years, technology has stepped in to support emotion regulation therapy. *Pocket Skills* was a mobile app developed in a 2018 research project to help people practice skills such as tolerating uncomfortable emotions while aided by an artificially intelligent electronic therapist, "Dr. Marsha" (named for the originator of one prominent therapy).[13] Those in the study had 24-hour access to reminders, education about emotion regulation, and could chat with "Dr. Marsha." The goal for those using the app was to learn and practice regaining emotional equilibrium after a triggering event, to avoid spiraling into extreme emotions, and to avoid making bad decisions based on their emotions.

Keep in mind that everything about this app had to be designed by human beings. Every decision about the content, interface, flow, tracking, and display all came from the designers' knowledge of human cognitive limits. The therapy app supported memory through reminders. It supported practice with motivating nudges. Of *Pocket Skills*, one research participant said, "Having something to explain [the skills] and refer back to in the moment anytime I want or need was an invaluable resource ... The app allows you to practice in the moment, every day, any skill and have quick access right at your fingertips, since you always have a phone." Researchers in Denmark tested a similar app with the same positive results.[14] These apps are just another example of how we can build the world around us to support our goals.

Learning to contain emotion in the moment is an admirable skill and one many strive to achieve. Even for people who have only experienced the momentary flashes of anger or fear that come with criticism rather than the extremes of a psychiatric diagnosis, it can seem desirable to quell emotions. But would it be better if we never allowed emotions to enter into our decisions?

Case studies from a few individuals unable to feel emotions illustrate that although we revere Mr. Spock, we don't actually want to be him. Take Damasio's case study of "Eliot," a man who had a tumor removed from his prefrontal cortex, causing slight damage to the area.[15]

Before the operation, doctors described Eliot as charming and successful. He had a family and a successful career as an attorney. The brain surgery was a resounding success and everything looked like it could go back to normal for Eliot. He scored highly on IQ tests and showed high ability in knowing all the things to consider in a decision. But after the surgery, Eliot was different. Gruesome or delightful pictures neither horrified nor pleased him. He declared he often felt like an outsider observing his own life, without any strong feelings about the outcomes. His emotions were muted, and he became unable to make decisions. It was as though he were aware of all the weights and possibilities, but unable to make the determination of what would be a good choice. One example Damasio described was Eliot sitting for hours trying to decide whether to organize files by name, date, or priority. Another time, Damasio asked Eliot where they should go to lunch. Eliot responded by generating all the pros and cons of each restaurant ("This one is usually empty, so there's no wait time. But maybe it's usually empty because the food is bad"). But he couldn't make a decision. All the things he wanted to consider needed to be weighed logically, an impossible task. Eliot needed emotions to cut through the weighting and help him focus on a few options (even if they weren't guaranteed to be the best).

Eliot was a fairly recent case study, but not the first person to experience life as Mr. Spock. Other patients with similar brain damage experience the same symptoms, from emotional detachment to lack of decision-making. As much as Mr. Spock may be idolized for his logical and scientific approach, it appears we need a bit of Dr. McCoy to function.

Of course, no one wants to be a purely emotional decision-maker. No reason or logic at all would make us less than human. We may be unconsciously affected, like the "hangry" (hunger = anger) judges mentioned earlier, but we still have some reasoning power. However, unless we have a particular type of brain injury, emotion sneaks into even our most logical decisions. So, how do we engineer ways to prevent or at least inform the decision-maker that he or she is about to pull an emotional Dr. McCoy instead of a logical Mr. Spock? One way is to take away the emotions we know are detrimental to decision-making, such as frustration or adrenaline, through design.

Dr. Spock?

Hospitals are a fertile area for introducing designs that encourage logical decision-making. Marc Resnick, a human factors psychologist, called out three areas medical designers should concentrate on to reduce frustration and improve decisions: (1) reduce the number of alarms, especially concurrent alarms, going off; (2) improve the usability of medical devices; and (3) improve the usability of the myriad systems workers consult, from software that only displays the first part of long medication names (this is a real thing) to making it easy to keep track of patient data for the "handoff" between shifts.[16]

Just to give some statistics and an example of the alarm problem, there are about *three hundred fifty* alarms going off daily for each patient receiving acute care in a hospital. Some have as high as seven hundred alarms going off daily. About 95 percent are "nuisance alarms" that don't need attention from the doctors or nurses.[17] It's not surprising when you think about the lack of coordination in the design of medical equipment – each machine has its own way of signaling, independent of the others. It might be good for the unimportant "I'm done" alarm for an IV bag to know to be quiet when another machine detects dangerous arrhythmias in a patient's heart, but no such systems exist.

Recently, I spent many hours visiting my mother in the hospital. I'm certain you could measure the spikes in our blood pressure when the alarms would go off, some for IV bags, others because it's easy to sweat off an electrode, many for reasons we never understood. In my case the heightened emotions from the alarms probably just made me more irritated at what was on the television, but it was frightening to think they might annoy the healthcare workers keeping my mother alive.[18] Across her fifteen-day hospital stay there was just a single "important" alarm that went off in the middle of the night at the nurses' station. We only became aware of it when, as my mother was asleep in the bed and my sister asleep on what passed for a couch, a nurse violently flung open the door and yelled "BREATHE!" This "alarm" worked brilliantly on both my mother and sister as they sat up and gasped in fright.

In sum, doctors and other highly trained professionals are not superhuman. Thinking that a doctor will always make the same decision, irrespective of alarms, unusable systems, and time demands is naïve. Instead, we can work to make these systems better, freeing the professionals to make decisions with a better balance of logic and emotion. We can do this by improving the usability of each system, considering the interacting systems holistically in design, and aiding those overloaded workers with checklists. Unburdened by the minute and frustrating interactions with technology, these workers should have their mental space freed up for real thinking and problem-solving. But once we have cleared the environment for good and creative decision-making, what can we do to improve the creativity coming from a team of people?

The Power of McCoy + Spock

It has been a robust finding that more diverse teams make better decisions.[19] Actively searching for people who bring different ideas to the table improves outcomes. Researchers have followed teams that were rated for diversity of race, gender, background, or experience. In study after study, diverse teams were more creative, better at solving complex problems, and better at turning a profit.

The best explanation for the benefits of diversity directly links to knowing, or thinking you know, the thoughts of the people around you. Professors Elizabeth Mannix and Margaret Neale led a massive review of studies on team diversity and concluded the penalty for non-diverse groups was that the group tended to assume everyone else thought like them (a call back to the discussion of Theory of Mind in Chapter 1). Professor Katherine Phillips, a later collaborator of Mannix and Neale's, summarized their findings in *Scientific American*: "Members of a homogeneous group rest somewhat assured that they will agree with one another; that they will understand one another's perspectives and beliefs; that they will be able to easily come to a consensus. But when members of a group notice that they are socially different from one another, they change their expectations."[20] Working from that assumption, these groups lacked the discussion and negotiation seen in the more diverse (and measurably more creative) groups. Exterior diversity (race and gender presentation) triggered interactions that showcased the ideas of their teammates because everyone on the team expected others would have different perspectives than their own.

Success is linked to interacting with those different from oneself, negotiating ideas and meaning, and coming up with more creative solutions. Great minds do *not* think alike. Emotional as he is,

Dr. McCoy acts brilliantly as Mr. Spock's interrogator and antagonist. Our predisposition to seek out those similar to ourselves may make decisions seem easier, but it lowers our creativity and the quality of those decisions.

Conclusion

Creativity, or the lack of it, is not predestined. Thinking can be infused with 'hybrid vigor' via diversity – from the diverse thinking of one person to diversity in a team. Emotion plays a large role in all creative and problem-solving situations, and it's dangerous to think that it doesn't. Recognizing when emotion is playing a larger role than it should means knowing what cues from our bodies and environment affect our decisions. We need to create technologies and systems that recognize these effects and work to un-bias us.

4 A MIND DIVIDED

> As every divided kingdom falls, so every mind divided between many studies confounds and saps itself.
>
> – Leonardo da Vinci

In 2005, Kira Hudson was in a car accident that left her paralyzed but energized to save others from the same fate. In her own words to the *Indianapolis Star* in 2020,

> When I was on the phone I was arguing. And it sounds crazy but I actually didn't even realize I wasn't paying attention to the road. I wasn't looking down at anything, I was just . . . my brain was focused on the conversation and not on the road in front of me. And then all of a sudden, looking up, there was actually a squirrel on the road in front of me and so I corrected but corrected too fast to the right, and then corrected to the left too quickly and then to the right which is what flipped the car.[1]

She called herself lucky. Lucky to be alive and lucky that she didn't hurt anyone else in the wreck. Lucky is what many people are, often unknowingly, depending on when it comes to escaping the dangers brought on by the distractions of technology.

You've probably intended to stop somewhere on your way home, only to find yourself at your own door, surprised at how you got there. While driving, you've had to react quickly to a deer in the road, or simultaneously judge the width of your lane, an oncoming car, and a bicyclist on the shoulder. Accidents are often related to these demands on attention amid distraction. The Virginia Tech Transportation Institute has been studying the effects of distractions while driving for years. They found that the chance of a crash increases threefold when using a phone; yet more than six-hundred thousand people use phones and other devices while driving every day.[2] What should be done? Ban the use of all cell phones and other distractions?

This question is more difficult than it seems. What is a distraction? Is talking to a passenger a distraction? Eating food from a drive-thru? Listening to the radio? Or is distraction reserved for interactive electronics, like cell phones and music players? What about a dash-mounted GPS? Would Kira Hudson still have had her accident if she had a hands-free cell phone?

David Strayer, Frank Drews, and William Johnston at the University of Utah have been investigating various distractions while driving since the dawn of cell phones, with some surprising results. The question they first asked seemed simple: How much does talking on a cell phone affect someone while driving? It was the early 2000s and as more and more phones came on the market, states hurried to pass laws against using them while driving. New York was the first, in 2001, with a ban on handheld phones. Headsets were still legal, which made intuitive sense because they didn't occupy the driver's hands.

This intuitive claim didn't sit well with the three psychologists. They began a series of experiments that pitted physical interaction with a handheld phone against mental interaction, either with a hands-free phone or talking with a passenger. They hypothesized that the limited resource tapped by cell phone use was attention, not the hands. After all, stick shift drivers have been moving their hands from the wheel since the first car was invented.

In a first experiment, Strayer and Johnston asked people to come in and respond to traffic signals in a driving simulator.[3] In the simulator, drivers had to do two things at one time. Some had to drive while talking on a cell phone while others had to drive while listening to the radio or audio books. Of those assigned to have a conversation, half used a handheld phone and half used a hands-free phone. All were told to discuss the same spicy topics – the then-current impeachment of President Clinton or the local scandal alleging bribery in the Salt Lake City Olympics. Those listening to books on tape were given a quiz to make sure they listened carefully, and only people who got *A*s on the quiz were included in the dataset.

The results were striking: the radio and books on tape groups were faster at completing the course and missed fewer traffic signals than *either* of the conversation groups. Further, there were no differences between the handheld and hands-free groups when it came to missing traffic signals: the groups were nearly identical and both had twice as many errors while driving as the group that just listened to the radio. So far, it was clear – using your ears wasn't much of a distraction, nor was using your hands. But was it talking that interfered with driving, or was it generating the thoughts that make for conversation?

It's Easier to Listen than Speak

The researchers again asked drivers to multitask. This time they asked drivers to either repeat words they heard on their earpiece or to play a game where they came up with a word that started with the last letter of the word they heard through their earpiece.[4] They also included a control group that just drove. This time, they also put them on driving courses varying in difficulty – either an easy course with few and very predictable turns, or a difficult, mountain road where they didn't know what was coming.

Everyone had more trouble staying in the middle of their lane on the difficult course. But what was surprising was how much worse those playing the word game did when the driving got harder: their errors went up more than either the control or repeat-the-word groups. Clearly, using attention to *generate* a conversation while driving had an impact on performance.

Frank Drews followed up these experiments by looking more closely at the kinds of conversation that would most hurt driving. In a study he led, people either spoke on a hands-free cell phone or to a passenger in the seat beside them.[5] Again, those on the cell phone did worse on every measure of driving – staying in their lane, following distance, and even failing to get where they wanted to go. But why? Fortunately, Drews recorded their conversations.

Those talking to a passenger tended to do two things that improved their driving:

1. When the driving got tough, like making a turn or braking, conversation drifted off, then resumed when driving got easier. This naturally happened for both drivers and passengers.

2. More of the conversation with a passenger was about driving and the driving context, with the passenger actually helping the driver notice their surroundings.

So, it was not conversing that hurt the cell phone users, it was that the person on the other end of the cell phone wasn't aware of the driver's conditions. Their incessant chatter used up the driver's limited amount of attention and sometimes there wasn't enough left for the road. Emotional or intense conversations, such as Kira Hudson arguing with her boyfriend, consumed even more attention. This brings me to an important caveat about the finding that passenger conversations don't have much impact on driving: for new teen drivers they do. And the damage increases with each passenger added to the car. Newer drivers still have to think a lot about driving, using precious attention, even while their passengers are likely just as excited, loud, and attention-grabbing as might be expected in teenagers.

Around a decade later, the US National Highway Traffic Safety Administration released a report of cell phone use on the road.[6] Over two hundred people allowed themselves to have their cell phones monitored while driving for about a month, including cameras in their cars that captured eye movement, lane keeping, and other indicators of safe driving. By this time it was more common to have cell phones integrated into the car itself, so they were able to compare handheld phones to hands-free phones, including hands-free phones that connected to the automobile. With their cameras and sensors they could tell if the car was in a crash, came near to having a crash, or if the driver had to take an aggressive evasive maneuver. Their findings mimicked the conclusions drawn from laboratory studies: simple conversation didn't increase risk. But all of the actions surrounding the conversation did, from starting and stopping calls to when the hands-free system failed and drivers turned back to their manual interactions with the phones: Handheld or hands-free didn't matter.

These experiments are good examples of how important it is to isolate an effect and establish a cause. If researchers had stopped when they saw the statistics on accidents and cell phones, we might still think crashes were due to drivers holding the phone. If they had stopped when they discovered it was the conversation that affected drivers, we might have laws against talking on the phone but also design changes that reduced in-automobile entertainment, like music, but for no valid reason. Or we might spend time trying to teach drivers not to talk to their passengers, when passenger conversations don't harm driving, sometimes they help it. Policy, law, design, and training are all affected by the results of such experiments.

Division of Resources

When multitasking, some tasks conflict more with others and some less. Sometimes we *can* do two things at once quite well. Radar operators can watch for a target on a screen and hit a button when a target appears at the same time they are answering questions via headsets. Many artists prefer to listen to music while drawing or painting. I spent many hours on the phone as a teenager, mindlessly playing Tetris while keeping up socially. What makes these task combinations seemingly immune to the constraints of attention we've discussed so far?

There are different pools of attention available for different kinds of tasks. In the 1970s, Alan Baddeley developed a model of attentional control that explained where you direct your attention, how much of it you use, and how long you can sustain it.[7] In Baddeley's model, there are three separate types of attention: one for vision and the location of things that need to be remembered, like displays and maps, one for words and things heard that need to be remembered, and a command center that controls the use of the other two. He called the resource for vision and location the "visuospatial sketchpad" and the resource for words the "phonological loop" (because we tend to repeat the words in our inner-ear to keep remembering them).

Driving, staying within the lane, and awareness of traffic requires vision and location information, so it draws some resources from the sketchpad and some from the central command. A cell phone conversation is an auditory task, and so uses resources from the phonological loop and some from the central command. If sensory demands change between the two tasks (another driver starts honking or the cell phone discussion involves complex ideas beyond "I'm fine" or "I'm almost there, see you in 10"), more of the command center resources are needed to balance each task with the driving. All the resources are finite, though, and you can't spend more attention than you have.

Baddeley's model was a great start for thinking about how attention wasn't just one thing. At nearly the same time as Baddeley's discoveries, human factors psychologist Chris Wickens expanded the idea of these attentional resources into Multiple Resource Theory (MRT), which can be used to roughly calculate how much conflict two tasks will have when performed together.[8] Thus, MRT turns the concept of separate attention resources into a method useful for design and other applications. In Wickens' own words, the purpose of MRT must be practical: "I felt it important that the dimensions of the model coincide with relatively straightforward decisions that a designer could make in configuring a task or work space to support multitask activities: Should one use keyboard or voice? Spoken words, tones, or text? Graphs or digits? Can one ask people to control an interface while engaged in visual search or memory rehearsal?"[9] MRT can help answer these questions.

Fuel for the Fire

The idea behind MRT is that we have multiple resources for attention, each with their own capacity, and it is only when we exceed that capacity on a task or tasks that our performance declines. The picture in Figure 4.1 is a simple matrix of the four containers of attentional resources. (This can be extended to include resources for smell, taste, and touch, but I'll explain it with only vision and hearing.)

Any task can be broken down into a combination of modes and codes. The whole of one's attention is like fuel, fuel which can be doled

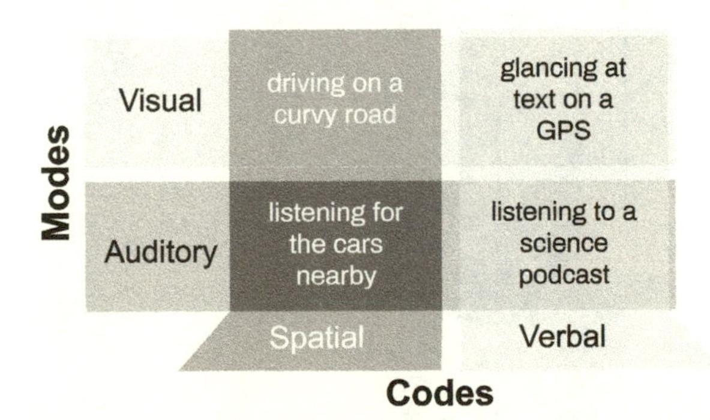

Figure 4.1 A layout for considering the attentional resources needed for a task or tasks. The demands each task place on attention can be categorized as a combination of incoming perceptual modes (whether they are seen or heard) and codes (whether they provide spatial or verbal information).

out when needed. You can't use more fuel than you have. Now, imagine that the fuel is divided into containers, each with a label for how it can be used. One container of fuel can be used for tasks dealing with words, like listening to a radio show; we would call this *verbal code fuel.* Dip into a different fuel container when you're faced with a spatial task, like driving on a curvy road. There are no words in this task, so it would not draw from the verbal code fuel, but would use spatial code fuel. Of course, you have to perceive each of these tasks, so they both get another label for their mode: visual or auditory. Radio shows are spoken, so listening is an auditory/verbal task (and would use auditory fuel and verbal fuel), but if you were reading the same information it would be a visual/verbal task. Driving a curvy road is visual/spatial, but you could imagine in-car automation that beeps if someone is in your blind spot. That would be an auditory/spatial task. Thus, unlike Baddeley's model of working memory where visual resources are always paired with spatial, here any combination of the two modes and two codes is possible.

Now, we can add another dimension to the resources we've already talked about: their stages of processing (Figure 4.2). There are

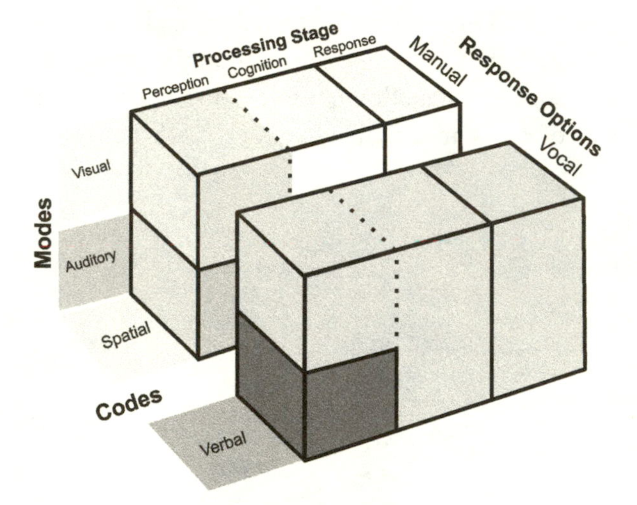

Figure 4.2 A full 3-D model of Multiple Resource Theory, showing the attentional resources that may be needed for a task or tasks, the stages of processing as initial information from the world becomes encoded in the brain, and a section for the resources needed to make a response, such as pressing a button or moving a mouse (manual) or providing assent or information verbally (vocal).

three stages. The first stage is perceptual: you need information and we can categorize that information into mode–code combinations. But we often need to interpret or use that information, not just perceive it. To do this, we move the perceptual information to a cognitive stage of processing. Take a close look at Figure 4.2 – the line dividing visual and auditory information stops at the cognitive stage. This is because once the brain encodes the incoming information, it no longer matters whether it was auditory or visual: reading a sentence is the same as hearing it once it becomes a memory. But the distinction between *spatial* and *verbal* remains – a reminder of Baddeley's model and the visuospatial sketchpad and phonological loop. The last stage, for many tasks, is the response stage, like turning a steering wheel, or saying "Ok, I got it."

Using the full MRT model, demand on attentional resources can be captured for any two tasks that will be performed together and the researcher can see how much they slurp fuel from the same tank. The more overlap (on modes, codes, stages, and response types), the more likely performance on the tasks will decline when they are done together.

Don't Try This at Home

Recently I was driving to a new city, following directions from the GPS in my phone, while listening to *Radiolab*, a wonderful science podcast. My triple-task load was a good example of resources and potential conflicts. Driving was mostly a visual/spatial task, with some visual/verbal attributes, such as reading the highway signs. Following the GPS was mainly an auditory/verbal task, as I waited to hear about upcoming turns, but also had some visual/verbal attributes because I often forgot what the GPS said and needed to glance down at the screen. I also sometimes looked at the screen to get an idea of upcoming turns so I could mentally prepare to follow the auditory directions. Last, my podcast, an auditory/verbal task, should have been the easiest and most dispensable – it certainly was not critical to driving. Yet, it was an engaging story that provided the most reward of the three.

In it, the hosts described an experiment on mice and stress that involved dropping mice into a tub of water and timing how long they tried to get out before giving up and floating. If you were listening to it with me, you would have heard,

> He dropped them in and as you'd expect, they try to escape. ... One minute passes. They swim, they swim! ... to the edge and all around looking for an escape. Two minutes pass. (Sounds of splashing in water) Three minutes pass ... (More sounds of splashing) but about four minutes in the mice start to get worn down, and then they decide at this point that there's no point, I'm giving up.[10]

As I heard that, I could *see* the struggles of that little mouse, the scrabbling and the splashing. The show grabbed my attention no matter what else was going on. The details were so visual that all the images conjured up in my head were using up some of my visual/spatial resources, the same resources I needed for driving.

The highest attention conflict should have been between the podcast and the spoken GPS directions, because they are both auditory and verbal. But my phone handled the conflict by prioritizing the directions: When they were given, the podcast faded out. This was irritating when there were many directions that I didn't really need, or long street names that interrupted my podcast, but reducing the conflict between directions and podcast was the right design decision.

Trouble only came when I needed to interact with the screen – although simple conversations on a handheld phone might not impact driving, looking through a menu to choose the next podcast certainly could. It was highly visual/spatial *and* visual/verbal – likely the reason I am safe is that nothing unexpected happened on the road when I switched from one podcast to the next.

There is one more element to consider when calculating task conflicts: estimate the level of demand on each resource.[11] This is a best guess at how demanding the task is of that resource and a typical scale is 1–3, where 3 means extremely high demand on this resource (continuous turning of a steering wheel with no chance to remove one's hands would be a level 3 demand of manual/spatial response resources). Or maybe it's just a level of 1. A boring podcast that my brain was already tuning out could be at that level for auditory/verbal and auditory/spatial. Anything that I've practiced enough to be automatic likely gets a 1. I've been driving for twenty-eight years now, so most of my steering and breaking gets a 1. The highest demand came from following directions and street signs, since the new city was unfamiliar. Once each resource demand is assessed for severity, this information can be added to whether there is conflict to predict success or failure in multitasking.

Change a Design *before* It's Used

MRT has been around a long time, but specific instances of how it changed a design are hard to find. This is probably because most industries are proprietary and don't share their processes with the world. BMW and Toyota aren't publishing how they studied and modeled performance with their dashboard controls and they certainly are not revealing any software they developed. Fortunately, one domain is open – government agencies. Researchers at NASA have been working on human performance modeling software since the early 1990s, calling it MIDAS, for *man–machine integration design and analysis system.*[12] MIDAS uses MRT as part of its predictions for how people will multitask (or fail at multitasking). You could feed in my story about driving, the GPS, and the science podcast and MIDAS could give you the likely outcome of what I would do if a cat ran across the road, or if a driver cut me off from the right lane.

MIDAS has gone through several iterations, each adding more helpful equations to guess at how humans would perform in a yet unbuilt or untested system. One of the earliest applications of MIDAS changed the design of nuclear reactor operation by Westinghouse.[13] When Westinghouse wanted to implement a new software system in their plants, the old paper-based procedures and new software system were put into MIDAS and run through imaginary emergency scenarios. Each scenario had little differences built in, such as who got what information, when they received the information, and how long it took each person to react.

In the paper-based system, MIDAS showed that the person in charge, the Senior Reactor Official, would be too tied up and distracted by the incoming questions from staff and requests for updates to do their job well. Clearly, the verbal channel was overloaded in all of the stages of processing, including responses. The cognition stage was also overloaded, because the Senior Reactor Official had to wait for requested information while doing other tasks. The attentional resources needed to keep the request active could not be used for the other tasks. This showed exactly what needed to be built into the new software system to fix these kinds of overload: provide the requests for information and the results in the interface, removing the need for the Senior Reactor Official to keep tabs on those tasks.

The modeled results weren't all sunshine for the new software system, though. The new computerized system was designed so that it unintentionally covered other important displays, which would go unnoticed in an emergency unless the operator remembered to go looking for them. Thus, using MIDAS early on pointed to important design changes before the system was even created.

All this was done with an early form of MIDAS. The newer versions have given it the ability to predict what might grab our attention, whether we want it to or not, and how tasks are done in teams compared to individuals.[13] Our minds are complex and we are still far from understanding them. But the fact that MIDAS and other models of human thought and performance are fairly accurate in predicting what real humans would do in a specific situation is an indicator of how far psychology and neuroscience have come in the last century.

Conclusion

Spaceships, cars, and nuclear reactors – they all need us to multitask at least some of the time. These are systems so complicated we'd have no hope of operating them without help. But we have a superpower, and that's the power to design the systems that never demand more than we can give.

When considering all the requirements for a new design, human factors psychologists and their collaborators have a multitude of considerations. First and foremost, they must consider the human operators with all of the attentional capabilities and limitations they possess. They must consider the tasks or tasks, with the kinds and combinations of mental and physical resources they demand. Lastly, they must consider the social milieu for the product or system: What social or cultural barriers must be understood or overcome? If cell phones in cars are here to stay, designs have to work with the fact that people won't stop talking on them. A mind can be divided, as long as it's divided in the right way.

5 ALL THE LIGHT WE CANNOT SEE

I am not absentminded. It is the presence of mind that makes me unaware of everything else.
– G. K. Chesterton

In the late spring of 1989, Lynn Hill, one of the most famous rock climbers of all time, tied into a rope for an easy warm-up climb. She had tied that knot thousands of times before, but this time, she was interrupted half-way. In a Q&A on *Reddit*, she said, "I went to get my shoes which were about twenty feet away, and I was talking to a visiting climber, and forgot that I didn't finish my knot."[1] She was wearing a jacket that covered the half-tied knot and also helped hold it in place as she began to climb. She easily reached the top of the cliff, with the knot still tucked into her harness. Ready to be lowered by her belayer to the ground, she leaned back on the rope and didn't feel the tension that indicated it would hold her weight. She pulled on the rope to tighten it and "instead the rope came out of my harness and I had no rope at all. And fell to the ground, me and the rope. Seventy-two feet to the ground." How could she have missed that crucial safety step, especially after the extensive practice she had over many years of climbing?

Pay attention. Give me your attention.
You're not listening to me.
I'm just so ADD.

You need to be more careful!

We take attention for granted. You seem to know if you're paying attention or not and whether others are paying attention to you. But what is attention? Can it be defined, measured, predicted, and controlled? How much control does a designer or advertiser have over capturing your attention? How much control do *you* have over your attention? I thought about these questions as I was driving to the North Carolina mountains while listening to *Dracula* on audiobook. I couldn't have told you what I passed, whether there were other cars, or what town I was near as Renfield ate his flies and Van Helsing stalked the vampire. Yet I was safe, on the road, between the lines, and thirty minutes closer to my destination. But where was my attention? And why hadn't I made any memories of my surroundings?

The limits of attention can cause both "highway hypnosis" and a half-tied climbing knot. Daniel Kahneman, psychologist and Nobel Prize-winner, called this seemingly mindless form of attention "System 1" and defined it as fast-acting, automatic, non-conscious, able to work in parallel with other processes, and requiring little effort.[2] One of the most important characteristics of System 1 is that it often just needs a "go" command to carry a task to completion. For example, the go command for Lynn Hill's knot would be "tie the knot." Once the tying has begun, most of the procedure was outside conscious attention: System 1 operates with little to no voluntary control. An unfortunate corollary of this is that when System 1 is interrupted, non-completion is not brought to the conscious mind, explaining the half-tied knot. To give System 1 a meaningful name for the rest of the chapter, I will call it the Automatic Processor.

The companion attentional system to our Automatic Processor is System 2, which operates slowly, serially, with full conscious awareness, and much greater feelings of effort. So much effort and control is needed for System 2 that it is often called "controlled processing" (which I also prefer as a label).[3] Kahneman noted that the Controlled Processor is often the system we most identify with, because its logic and conscious decisions are within our awareness and feel like part of "us." Following the plot of a movie or book, learning geometry, deciding where to submit your resume or where to live all use controlled processing. Without it, you wouldn't be you. However, more often than not, the *Automatic* Processor is in control of our behavior. This is usually a good thing, or we would spend our days thinking through how to put on clothes, how to pour milk, and driving like we just got our licenses.

Don't Take My Word for It

For a quick demonstration of Automatic versus Controlled Processing, try the following: without using any cheating strategies, such as covering part of the words or blurring your eyes, quickly name the animals in Figure 5.1 *without* reading the words on top of them.

Did it feel effortful to *not* read the words as you named the animals? Your Automatic Processor already read the word "elephant" even as you deployed the Controlled Processor to recognize a dog. Thus, your Controlled Processor had two tasks: first, suppress Automatic Processing from saying "elephant" and then *also* recognize and name the picture of a dog. The two attention systems are in opposition for these kinds of tasks, making it feel harder and take longer than doing either task by itself.[4]

Of the two processors, your Controlled Processor is more affected by alcohol, far more than your Automatic Processor. Folk knowledge is correct that alcohol reduces inhibition; indeed it does, by reducing your ability to inhibit automatic responses. That drunk confession by a friend? He or she was thinking it, but without the alcohol, their Controlled Processor could have stopped them from blurting it out. Alcohol is not alone in this: there are many substances and situations that reduce our control over automatic processing by controlled processing, including fatigue, older age, and conditions such as attention-deficit disorder (ADD) or attention-deficit and hyperactivity disorder (ADHD).

It's Not an Attention Deficit

The word *deficit* is right in the label, but it is not an amount of attention that is lacking for those with ADD or ADHD. Rather, they

Figure 5.1 Name the animals, but don't read the words.

have trouble using their Controlled Processor to stem the flood of information and decisions coming in through their senses that activate their Automatic Processor. They can seem restless, excitable, or impulsive – traits usually kept in check by a Controlled Processor.

This may make it sound as though the Controlled Processor is more important, and that we suffer without it. However, although we more often identify with the conscious control over our thoughts, we should not malign automatic processing: it is related to other attributes we value, such as creativity, empathy, and imagination. Surprising connections, those "aha!" moments, between disparate ideas, are brought into consciousness from the Automatic Processor. As Steve Jobs, of Apple fame, said to *WIRED* magazine, "Creativity is just connecting things. When you ask creative people how they did something, they feel a little guilty because they didn't really do it, they just saw something. It seemed obvious to them after a while."[5] The Controlled Processor still needs to make these connections conscious, but they were whispered to it by the Automatic Processor.

Barry Kaufmann, a researcher at the University of Pennsylvania's Imagination Institute, is an outspoken champion of our Automatic Processors, demanding they receive the credit they deserve for our creativity, imagination, and aesthetic sense. He says those with ADHD have a "leaky filter" that allows even more connections between seemingly disconnected ideas.[6] "By automatically treating ADHD characteristics as a disability – as we so often do in an educational context – we are unnecessarily letting too many competent and creative kids fall through the cracks," said Kaufmann. "It seems that the key to creative cognition is opening up the floodgates and letting in as much information as possible. Because you never know: sometimes the most bizarre associations can turn into the most productively creative ideas."[7] Maybe a little leaking is a good thing.

Our Automatic and Controlled Processors work in tandem, balancing the benefits of logic and intuition. Most of the time they do a great job, although we can have large differences between people in terms of how well their processors work together. But even for the most logical, even those who value the performance of their Controlled Processor the most, the Automatic Processor is often more in charge of their decisions than they would like to believe.

Thirty-Seven Children a Year

On a sunny, hot day in July, Kyle Seitz was driving his 15-month-old son, Benjamin, to daycare. This was all part of the routine; a routine that involved the drop-off, grabbing a coffee, and heading to work as a software developer. Some events on that morning were not routine: It was the end of a holiday weekend, the fourth of July, and Seitz's in-laws were visiting. Kyle changed Benjamin's diaper, gathered up the daycare supplies, picked up the infant, carefully buckled him into his DOT-approved, backwards-facing car seat, and drove off.

We'll never know what caused his attention to blink, but Seitz found himself at the coffee shop, ordered his coffee, and left for work with Benjamin still strapped in the car seat, probably asleep. Court records would later show, as they were deciding whether to charge Seitz with murder, that he got back in the car to grab a sandwich for lunch and, in the evening, drove to the daycare to pick up his son.

Seitz's lack of memory for the morning drop-off was so complete, he told police that he hadn't been alarmed not to find Benjamin at the daycare because it seemed likely his wife had already picked him up. He only became worried when they said Benjamin hadn't been there all day, at which point he ran back out to his car. In the warrant for his arrest, it was noted that "Kyle Seitz stated that at no point during the course of July 7, 2014, from leaving his house to being told that (Benjamin) was not at daycare did Kyle have any sense or inkling that (Benjamin) remained in his vehicle or was anywhere else other than safely at daycare." His wife stated that she wondered, when she heard her son hadn't been dropped off at daycare, if "this stuff you hear about every summer happened to us," and later, "we all think it can't happen to us."[8]

Public opinion on the case was clear and the urge to blame the victim was strong: *everyone* thinks it cannot happen to them. Comments on news stories for this and the approximately forty deaths like it every year are always the same, a variation of *I can't imagine forgetting my child was in the car. There must be something wrong with someone who does this.*

Blaming the victim serves our need to believe we are in control and thus that such a tragedy could not happen to us. But since our Controlled Processor is in charge of "being careful," and it is severely limited in *everyone*, we are all "the Seitzes." If the right circumstances lined up to deplete our controlled processing resources and activate automatic processing, we would also not be conscious of our mistakes and thus be unable to avoid them.

From Tragedy to Policy

This case, and the 500+ like it from 2000 to 2015, shows how easily we can lose control of our attention. The Seitzes have been working to have sensors installed in cars to prevent such accidents but have not been successful (as of 2021). In an interview with CNN two months after the incident, Lindsey Rogers-Seitz showed knowledge borne of tragedy, saying that safety technology should be fitted as standard, not as an accessory, because people do not understand that such a lapse of attention could ever happen to them. She said she herself probably wouldn't have purchased the safety feature on its own, never expecting that she or her family could forget their child in the car.[9]

Unfortunately, the only other members of the public who seem to understand and believe this are those who have shared the Seitzes' experience (and, of course, research psychologists). As CNN reporter Sunny Hostin said in a confessional news story about her own baby,

> We parked, turned off the ignition, closed the sunroof tight and locked the doors. I went to get a shopping cart while my husband stood by the car waiting. We then walked together past the car and toward the crowded Home Depot. Walking into the garden center, my husband turned to me and said, "My God. We left Paloma in the car." ... Before I left Paloma in that hot car ... I would have said that you must prosecute even those parents who leave their kids in a hot car accidentally because you must send a message to parents that they have an obligation to protect their children. Before I left Paloma in that hot car, I would have said that the anguish that this parent would feel is not enough of a punishment. Had you asked me about one of these cases, I would have told you that I am a watchful mother and would never, ever do that. But I left her. My husband did, too.[10]

As Hostin said, it's hard for anyone to believe such a life-changing attentional blip can happen to them, leading to blaming the victim. In opposition to this, the public response to the story of Lynn Hill that started this chapter was telling. I have frequently heard in the rock-climbing community that "if it can happen to Lynn Hill, it can happen to anyone," the opposite of blaming the victim. Lynn Hill was considered a true expert, skilled, experienced, and well-respected. Yet, as experienced and famous as Lynn Hill was, she was not more experienced at climbing than Kyle Seitz was at driving to work every day. Both fell prey to their Automatic Processors.

Aside from being a good reason not to create designs based on public opinion, there is much to learn about designing for human attentional limits from these stories. Clearly, a safety feature is needed, but what form should it take? Perhaps it should be an electronic sensor in the car, or built in to the car seat. Perhaps a camera that faces the car seat, with face-recognition software? A cell phone app, paired to a device on the child? The technology could take many forms, and once imagined it's important to test them for failures: What if that sensor failed? Tell the user? Disallow the car from starting? Lock the door? Sound an alarm? How redundant or cumbersome a system will users accept when the tragic occurrence is so serious, but also so rare? What privacy concerns might keep parents from welcoming certain safety features? As with all product ideas, a good beginning is to start by considering human attention and its limitations.

Can We Do Better than an 11-Year-Old?

We learned earlier that there are two attentional systems at work, automatic and controlled, and that the Automatic Processor will take over whenever possible. The drive to work, the daycare drop-off, and trips to the grocery are likely all automatic processes. We know that when an automatic process is interrupted, the mind does not always consciously note that the task is incomplete or which sub-task remains undone (such as Lynn Hill's untied knot). Thus, it's a good assumption that the design needs to recognize the interruption of a routine task and do what the mind will likely fail to do: alert the user of the interruption.

Of course, sometimes a child might be left in a hot car *because* of an uninterrupted routine task. For example, a parent who never does daycare drop-off suddenly needs to (but continues on to work like every other day), or a parent takes the child on an errand where the child is never along (and accidentally leaves them in the car while shopping).

Figure 5.2 Recreation of the memory aid developed by Andrew Pelham at 11 years old to prevent leaving toddlers in the back seat. Brightly colored elastic attaches from a point near the child's seat to across the driver's door, blocking exit. Keeps dogs safe, too.

Our automatic processing system is still at fault, but this time the Controlled Processor must *interrupt* the task and remind the parent to get the child from the backseat, a job it's bad at doing.

One potential solution was developed by Andrew Pelham, an 11-year old, as part of the Rubber Band Contest for Young Inventors.[11] The "EZ Baby Saver" is a brightly colored string of rubber bands (natch) attached near the child's seat and clips to the interior of the driver-side door (Figure 5.2). Thus, it physically interrupts the driver any time they exit the car. However, there are potential downsides, not the least of which being that it requires action from the driver to be put in place each time, and as we know, people do not believe they could forget their child. It also might be difficult to enter the car with it in place, or difficult to fasten once in the car. Lastly, some people put the baby seat in the passenger-side rear, meaning the alert on the driver side would not be located near to putting the child in the seat, a violation of the *proximity compatibility principle*, where displays and controls that depend on each other should be located near each other.[12] Despite these design criticisms, please note it accomplishes the important cognitive task of interrupting any automatic process and prompting the necessary controlled one to keep a child safe. Designers can take this same concept and create products that meet the same goal, with anything from a built-in sensor to a smartphone application. That said, if a solution were available, would anyone buy it?

"Safety Doesn't Sell"

There's a particular type of injury common to woodworkers and builders. Imagine a hand, held up in greeting. It is an incomplete hand, with multiple portions of fingers missing, typically in a diagonal formation. The culprit is the table saw, one of the most versatile tools in construction. The most common time for it to happen is when a woodworker is pushing material into the saw for a cut, resulting in about 30,000 injuries each year and more than 3,000 amputations.[13]

Like every skill discussed thus far, woodworking is a combination of automatic and controlled processing, with an emphasis on the controlled processing. There can be a large number of interacting variables in cutting wood, such as precise measurement, wood density and amount of pressure needed to push it through the saw, knots, or other imperfections that interfere with the cut, and dampness of the wood. Every woodworker values their fingers, but it's no surprise that with a controlled processor caught up with so many other goals and decisions, hand position gets lost.

The *SawStop* company went on a mission to take away the punishment of an attentional lapse. Their solution was an engineering one – if skin makes contact with the blade, the conductance of the skin sends a signal to a brake that instantly stops the spinning saw. Videos are readily available online, and I recommend watching. The speed at which the blade stops is truly incredible. That said, I wouldn't have wanted to be the first person to test it with my hand, even after seeing some convincing demonstrations with hot dogs, and even though the National Consumers League said, "The evidence is overwhelming that this very dangerous product can be made almost entirely safe."[14]

So, should the *SawStop* (or similar technology) be required for every high school workshop? Should it be standard on all new table saws? As soon as that thought was out there, there was pushback. The arguments were that a *SawStop* would teach students to be careless with table saws, and then have more accidents when they don't have a *SawStop* available. The Power Tool Institute, an industry group, was quoted as saying the *SawStop* would increase accident rates via a "false sense of security," albeit without data to back up the claim.[15] There are two principles that apply, though. First, novices at woodworking are the most likely to have accidents. Their attentional capacity is full – full of all the new numbers, measurements, and concentration on how to hold the wood and move it smoothly. Perhaps they feel the pressure of being watched by a teacher and their peers. These demands are multiplicative, because the movements of the hands affect the speed of the machine, interacting with the density of the wood and other variables. Once these students become experts, with their own wood shops and thousands of hours behind them, they may still have accidents, but those accidents are more like Lynn Hill's fall – not due to an overfilling of capacity, but due to an automatic process being interrupted, or running to completion despite the need to halt the process before an accident occurs. It is not because they aren't "careful."

I argue that without a prevention such as the *SawStop*, those accidents would occur no matter what technology they first learned to use. But as long as those with all ten fingers still believe they are "careful" enough and fast enough, it's hard to convince them the extra cost of mandatory safety features is worth it. The same is true of all the parents who can't imagine forgetting their child in the car.

Despite other manufacturers refusing to license *SawStop* technology, it does appear the company had an influence. At least two other similar innovations have been developed – both taking away the severity of a momentary slip of attention with a table saw. As of 2021, none of these were standard or mandatory.

Conclusion

In 2019 I took an exam for certification as a rock-climbing instructor. Across two days I had to demonstrate knowledge of techniques, systems, and, of course, safety. As the first example in this chapter illustrated, rock climbing is an inherently dangerous sport. But the exam was not only on technical skills, it also covered the skill of teaching. We were told to come up with a topic and teach everyone else taking the exam. The examiner assessed us for how well we taught. It's typical that examinees teach a skill as they would to a new climber, such as how to belay or build a strong anchor that secures climbers to the rock. I chose to teach about the futility of "being careful." I could tell I got the attention of the examiner when I led off with "I'm going to teach you to avoid trying to be careful." His eyes flashed with alarm, as the mantra of life-endangering sports is *to be careful.* But by careful, we often mean to devote attention to what we are doing as it happens, and as every example in this chapter has shown, this is often impossible. I then said, "Everyone tells you to be careful, and that means attentive, but we aren't always in control of our attention. Distractions grab attention, multi-tasking uses it up, and we don't start off with a lot of attention in the first place." I showed them photos similar to Figure 5.1 in this chapter, only instead of animals and incorrect names I included climbing knots. It was fun for everyone to see how badly they did when trying to name a knot they knew well but instead read the wrong name as it flashed by.

> This is just one tiny example of how you can't control your attention. You want to name the knot. You're trying not to read the words. But you don't have the control and, at best, it slows you down. At worst, you name it wrong. This is the same reason that people fall of the ends of their ropes, don't finish tying their knots correctly, drop their friends when belaying, and forget gear that they need. Losing attention can mean dying or killing someone else.

Admittedly, I used more technical language than in this re-creation, but you get the idea. If they depended on being careful, at some point they would fail. And whether that failure meant nothing or meant death would be a matter of luck.

> I'm showing you how you can't be careful. The only way to protect yourself is to invest in systems that will protect you when you inevitably get distracted. For example, always tying a knot in the end of a rope means that you don't *have* to be careful or use attention to monitor where the end of the rope is. If you get to the knot, you don't end up falling off the end or dropping the person you're belaying.

None of the information I was giving them was new – they knew they should tie knots in the ends of their ropes, for example. But what I think was new was having to acknowledge that these systems aren't a backup for them being careful. Being careful isn't a viable safety technique on its own, ever. And, maybe most importantly, they perhaps understood that they shouldn't castigate themselves for forgetting or being distracted. They should put systems in place before they start, using their Controlled Processor so that *when* (not if) they are distracted, disaster is averted.

Even people with the highest attentional control, the best humanity can hope for, are not in full control. We need systems that work with our limitations, supporting our constrained attention. We also need an educated design force, knowledgeable about these limits and able to create technology that doesn't ask more than we have to give. Lastly, we need to be demanding consumers – not blaming ourselves when frustrating designs overload us and purchasing products with usable designs. Practicing cutting wood and "being careful" will only take you so far – you can't improve your working memory or attention (at least, not enough to matter), so it is technology and systems that need to make up the difference to keep us safe.

6 MISTAKES
We've Made a Few

If others had not been foolish, we should be so.
– William Blake, 1790, *Proverbs of Hell*

Unreal City

In 2011, I woke up to the news that an earthquake and tsunami had hit Japan. Aerial video of Miyako, one of the hardest-hit towns, looked like the Gulf Coast where I grew up. Ships were moored near shady pavilions, houses sat on stilts with docks and yards that were half-sand, half-grass. Inland, the town looked more like San Francisco, with white concrete streets, tangles of utility lines, and townhomes with views of the rocky coast. Even further ashore were farms with neat borders and stucco houses. My first sight of this sleepy looking town was to watch an inexorable flow of opaque water push its way over the sea barriers and into the streets.

Watching the wave from the air was both unreal and the most real. The cameras followed the slow-looking wave from far away, but I could see how fast the water was moving by the cars unable to escape it. I expected to see some resistance from the larger buildings, or at least some hesitation when the water hit them, but they were moved and tumbled just as easily as the cars. Many of the buildings were on fire, even as they floated on the wave. In one clip, a white car stopped, seeming to realize the situation, and turned around to retreat. Seen from the helicopter, the black waves neared, moving like a spilled drink on carpet. The white car looked like it just might make it when the tip of the wave tumbled them off the highway with the rest of the debris. That image in particular I cannot get out of my head. This was the backdrop for the coming of three nuclear meltdowns at the Fukushima nuclear plant.

Withstanding Mother Nature

Earthquakes and tsunamis are "natural" disasters, but much of their impact depends on human planning. The earthquake that triggered the tsunami was 9.0 on the Richter scale, the second strongest quake in recorded human history. Yet the high-rise buildings of the Japanese coast survived. The buildings that were damaged by the quake were old, built without the shock protection of modern structures. Footage of the earthquake included numerous office buildings, with walls swaying like reeds, but the only damage came from furniture and file cabinets tumbling within. Engineers thought they had accounted for tsunamis as well, but in Fukushima Daiichi, the seawall protecting the nuclear power plant was only built to withstand a wave half the size of the 2011 tsunami.[1]

In the following hours and days, errors compounded at all levels, reaching to the upper echelons of the Japanese government and the company, TEPCO, that managed the nuclear plants. As one TEPCO employee of the plant said in an interview with PBS *Frontline*, "The more you learn about the nuclear power plant, the more you see the complexity of the systems. There are backups, which also have their backups. We never thought that all of those double or triple backup systems would become unusable." Other workers echoed this line of thought:

> At that time, we could not imagine that the tsunami would sweep away heavy oil tanks, equipment and pumps to cool down the nuclear reactors; that it would flow down to the basement of the building and submerge the diesel generator which was equipped for the emergency backup; and that it would get all types of the power sources and the pumps damaged. Something beyond our scope of our assumptions occurred.[2]

Accidents of this magnitude, whether in nuclear power, aviation, medicine, or other life-and-death domains, are caused by many issues – but people are always involved. Analyzing the human factor in these accidents is our best hope at preventing them.

Good Cheese Is Solid

I can't write about major accidents without writing about cheese. Swiss cheese. Swiss cheese explains a variety of disasters – as well as how to prevent them.

The appropriately named James Reason, a British psychologist born in 1938, came up with the Swiss cheese analogy to describe the seeming coincidences that spawned horrific accidents.[3] The Swiss cheese model is a list of the levels where errors occur, ranging from the physical and mental state of a single person to the culture of an entire industry. Each slice of cheese should be a barrier, but sometimes there are holes in those defenses. In major accidents, such as nuclear meltdowns, airplane crashes, and the loss of the *Challenger* space shuttle, errors at each level lined up, like holes in slices of Swiss cheese, and allowed the accident to occur (Figure 6.1).

To understand the Fukushima meltdown and how to analyze a major accident, we can turn to the most analyzed accident in history: Chernobyl.[4] Books, movies, and TV series have documented every minute and interviewed every person connected to the disaster – the

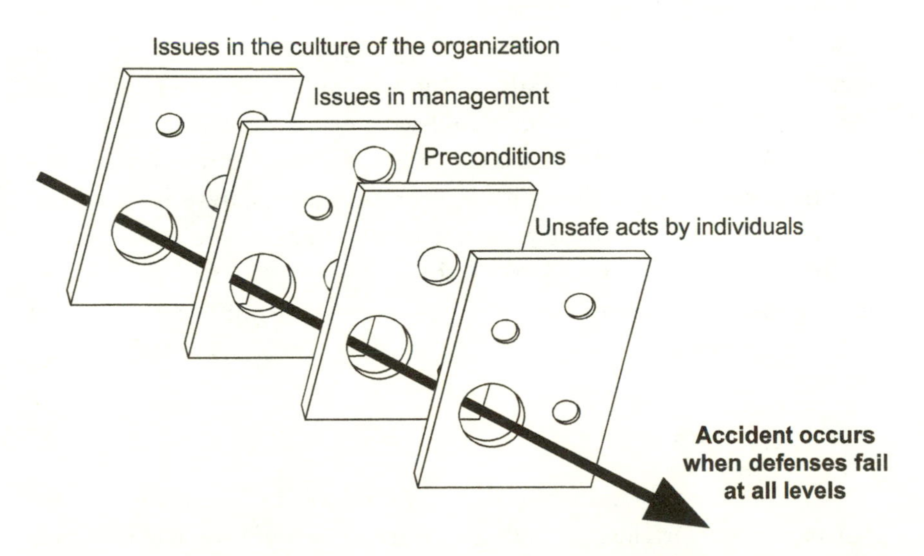

Figure 6.1 The layers of human systems and technology where failures at all levels can invite large-scale tragedy.

ones who lived, at any rate. It turned out to be a classic accident, showing every hole in the Russian cheese.

Chernobyl was not the first nuclear accident on Russian soil. A multitude of smaller accidents prior to Chernobyl were successfully covered up by the government. But as radioactive fallout blew into Sweden, the USSR could not hide Chernobyl. The level of historical analysis and detail available allows vivid descriptions of each layer of Swiss cheese, from the actions of the workers to the preconditions that set the stage for the accident, all the way up to the supervisory and organizational failures in the government nuclear system.

Say No to Victims

Science requires victims.

– A. M. Petrosyants, Chairman of the Soviet State Committee on Utilization of Atomic Energy, 1986 press conference on Chernobyl

If there were ever a statement in direct opposition to everything human factors psychology stands for, it was the claim, "Science requires victims," coldly stated by a Soviet official after the Chernobyl disaster. The official was a mechanical engineer with no scientific reputation when appointed to run the Soviet nuclear program. Perhaps it could be rephrased as, "Poorly designed or managed systems eventually take victims." The more one reads about Chernobyl, the more it seems destined that the Soviet Union would see a nuclear meltdown.

We'll start at the last layer of Swiss cheese, at the unsafe acts that triggered the explosion. This is where accident analysis used to start and stop, blaming the single person who made a mistake. Now accident analysts explore influences beyond the moment of the accident because finding the hidden triggers in the other layers is the only way to prevent future incidents.

Unsafe Acts

The cheese level we most naturally look to for blame is called "unsafe acts." Driving through a red light is an unsafe act. Letting a toddler play unsupervised around a pool is an unsafe act. Pointing a gun at someone for fun is an unsafe act. We're quick to lay blame at this level, on the inattentive driver, the neglectful parent, or the foolish gun-holder. These unsafe acts are the "active" errors and often the most observable. In a radical rephrasing of these errors, Sidney Dekker, a professor and accident analyst in Australia, proposed that we consider these errors the outcomes of a failed *system*, not the error of any one person.[5]

The *unsafe act* that caused the Chernobyl meltdown was dipping control rods quickly into an already too-hot reactor. A quick primer on nuclear reactions: radioactive fuel rods are kept in check by control rods that can be inserted or withdrawn to keep the reactor at a safe level. Without the control rods, the fuel rods produce a runaway reaction where the reactor gets so hot the fuel rods melt through the bottom of the reactor, releasing radioactivity – a "meltdown" (and a steam explosion that spreads the radioactive fallout).

At Chernobyl, the control rods were inserted while the reactor was still in a controllable state. However, the rising heat alarmed the lead engineer, Aleksandr Akimov, who hit the button that dropped the control rods fully into the reactor in an attempt to stop the nuclear reaction. When the rods entered the hot core of the reactor, they exploded, bringing down parts of the building and the safety systems designed to limit the nuclear reaction. Instead of stopping the reaction, it heightened it into meltdown.

But why was the reactor hot enough to need the control rods reintroduced? And why were they taken out in the first place? This is where we move beyond the single unsafe act of Akimov. It turns out that an experiment was being conducted that night at Chernobyl, an experiment that opened up holes in the defenses of the power plant and precipitated Akimov's actions.

The experiment was to see if the reactor could keep spinning when the plant lost power and it involved cutting all electricity to the reactor. This may sound like a crazy thing to test, but Soviet reactors had been under attack before, and the government wanted to know how to avoid the very meltdown that they ended up creating. And so, they cut the power and deactivated the reactor safety systems.

Despite the bad design of the experiment, without the other issues at Chernobyl and the dismantling of safety systems, everything probably would have been fine.

But Aleksandr Akimov pulled the switch that dropped the rods. Wasn't he responsible for the meltdown? When we ask how could this happen and how can we prevent it, it is rare that the answer is the person who committed the unsafe act. Unfortunately, blame is what we most desire – as though, if we can excise that one bad apple, we'll all be safe again. But the real source of problems can't be traced back to a single nefarious operator. Indeed, in the current example the poorly planned experiment already had preconditions that foretold disaster.

Preconditions

The next cheese slice up from the unsafe act contains the *pre-conditions* that set the stage. For Chernobyl, there were three: the training of the workers, the design of the plant, and the design of the interfaces controlling the plant.

Nuclear plants in Soviet Russia often selected workers that were most loyal to the party rather than for having the best minds or accomplishments. Once hired, workers avoided speaking out or dissenting. And there were many reasons to cherish a job at Chernobyl: unlike many citizens, the plant workers were well taken care of and lived in a town with good resources and education for their families. Workers not only feared losing their job, but being sent to prison or Siberia for disobeying orders. Losing one's job at Chernobyl meant a lot more than just finding another.

Design of the plant was also unsafe. In the *unsafe acts* section, I mentioned the control rods being the cause of the explosion. Without going into engineering details, it was known that the graphite tips on the rods were a flawed design. The need to update these rods was common knowledge in the global dialogue on nuclear safety.

As to the interfaces in the plant, the safety systems allowed for manual override of the safety guards. Most of the safety systems were overridden to carry out the power failure test. One of the overridden systems was meant to prevent the reactor from generating too little power, a counterintuitive danger. Too little power allowed the reactor to be put into an unstable state because of "poisonous" byproducts, so that any number of *unsafe acts* could cause an accident. There were alarms, but nothing to prevent operators from choosing a too-low power output or overriding safety systems designed to kick in when they sensed danger.

But even dangerous preconditions don't guarantee an accident. If nuclear workers are tired, but there is another worker to take over, then the precondition of fatigue is acceptable. If workers are not the best, they will be fine as long as the job doesn't demand more knowledge or experience than they possess. Further, with the right supervision and safety systems in place, there is little chance for any individual worker to make a huge mistake. This brings us to the next level: unsafe supervision. The supervision, if one could call it that, was abusive at Chernobyl.

Unsafe Supervision

> You goddam idiots, you haven't a clue! You've screwed everything up, you boobs! You're ruining the experiment! I can't believe what a bunch of assholes you are!
>
> – Anatoly Dyatlov, 1989, Vice Chief Engineer, Chernobyl, quoted in *The truth about Chernobyl*

The supervisor with the most disparity between his knowledge, abilities, and job description may have been Anatoly Dyatlov, the Vice Chief Engineer at Chernobyl and in charge the night of the accident. Grigori Medvedev, a nuclear engineer and investigator of the accident, said that when Dyatlov was hired he was inexperienced and "not acceptable for the post of chief of the reactor section. He would encounter difficulties as manager of the operators not only on account of his personality and his obvious shortcomings in the art of communication, but also of his previous work experience, since he was a pure physicist with no knowledge of nuclear technology." That was one of the nicer quotes available about Dyatlov, as Medvedev also called him "slow-witted, quarrelsome, and difficult," and Razim Davletbayev, deputy chief of the turbine unit in the No. 4 reactor said,

> He did not work particularly hard. He rejected all proposals and objections that required any effort on his part. He was not involved with the training of the operators . . . He had to spend a long time pondering the substance of questions, although he was quite a capable engineer. With him in charge, people worked without any feeling of satisfaction. He was a stubborn, tedious man, who did not keep his word.

These quotes were from interviews contained in Medvedev's investigative report, *The truth about Chernobyl*, which won the *LA Times* Book Prize in 1991.

Lastly, the Director of Chernobyl was asleep at home during the doomed test and subsequent disaster. He was known for saying that it was more likely one would be hit by a meteor than have a nuclear accident at his plant. In this case, his unsafe supervision was no supervision at all.

It would be tempting to heap the blame on Dyatlov and the other under-trained workers for unsafely running an already dangerous experiment. It is also tempting to blame him because he was so unlikable (although character and competence are not related). But Dyatlov was only a piece of the system, and if he were the only hole in the Swiss cheese, his unsafe supervision would likely have caused no harm.

Organizational Factors

> There you have our whole national tragedy in a nutshell. We ourselves tell lies, and we teach our subordinates to lie. Lies, even for a worthy cause, are still lies. And no good will come of it.
>
> – R. G. Khenokh, 1989, senior construction officer of the Zaporozhiye nuclear power station, USSR, quoted in *The truth about Chernobyl*

Lastly, the most often ignored (but also the most critical) layer contains the cultural and organizational factors that contribute to accidents. Chernobyl was isolated from international nuclear sites and from their common knowledge base. Nuclear engineers collaborate across borders, sharing knowledge and experiences to improve the safety of reactors. The reactors of Soviet Russia received no collaborative help. Soviet distrust of others forced their scientists and engineers to work alone, under the command of an electric company that didn't know how to run a nuclear site.

The Soviet government isolated and lied to scientists, engineers, and the public for almost forty years about the number of small accidents and near misses in their nuclear sites. Even after the accident, the government tried to minimize the disaster in the same way a child points at another to say he or she did worse. This was the press release on April 28, two days after the meltdown: "The accident at the Chernobyl Atomic Power Station is the first one in the Soviet Union. Similar accidents happened on several occasions in other countries. In the United states 2300 accidents, breakdowns, and other faults were registered in 1979 alone."[6] Of course it was *not* the first nuclear accident in the Soviet Union, it was just the first anyone knew about. But such propaganda made the nuclear workers complacent – if there were no accidents, the systems must be infallible. An actual gag order was signed a year earlier by the Minister of Energy: "Information about the unfavorable ecological impact of energy-related facilities (the effect of electromagnetic fields, irradiation, contamination of air, water, and soil) on operational personnel, the population, and the environment shall not be reported openly in the press or broadcast on radio or television."[7] The damage from such willful obfuscation cannot be overstated.

Chernobyl provides an illustrative trip through the holes in Reason's Swiss cheese, with egregious failures at every level from the national culture to the last decision of Aleksandr Akimov. Decades later, a similar lesson came from Fukushima. The disaster *preconditions* were set with the inadequate design of the weather protection, but were exacerbated into disaster by *organizational factors*. TEPCO and the Japanese government did not have the same information as the "boots on the ground" workers, yet their hierarchical structure required decisions to come from the top. Workers waited on a decision to cool one reactor with seawater, which would destroy it for future use, resulting in a meltdown. When the Japanese government stepped in, their information came in a game of telephone from the workers at the plant, through TEPCO executives, and finally into the ears of the non-nuclear-savvy politicians, such as Prime Minister Naoto Kan. Decisions couldn't be made quickly and were not based on solid information: trucks were sent and could not get to the plants due to the conditions of the roads after the earthquake and tsunami. Workers were afraid of giving any information they were not entirely sure of, such as that the coolant in the reactor was gone, because they had gotten in trouble in the past when they had been wrong.[8] The holes in the cheese lined up, and the tsunami wave roared right through them. Fukushima proved that it is not enough to analyze previous accidents. We must make conscious efforts to apply the lessons we learned in every new complex system.

It's Just a Fad

Chernobyl and Fukushima captured the world's attention, and with good reason. Other accidents, particularly graphic ones, also capture attention but sometimes this can make us focus on the wrong safety issues. One example would be the Germanwings (owned by Lufthansa) pilot who, in 2015, deliberately flew a plane with 180 people into the side of a mountain to commit suicide.[9] This suicide/mass murder was an emotional, memorable story, making us think that we need extreme safeguards in case pilots try to crash planes with us in it. But just because something is scary and easy to remember doesn't mean it's the biggest safety problem out there. But the public tends to focus on these memorable events, crying out for a solution to *that* problem.

These flashbulb memories for horrific events end up in our laws and policies. The outcry over the Germanwings crash led to multiple countries adopting rules that two crew members must be in the cockpit at all times.[10] Other suggestions, particularly in the media, were that there should be a loosening of patient–doctor confidentiality regarding mental health issues of pilots, which could potentially impact the privacy of hundreds of thousands of airplane pilots worldwide. The two-in-the-cockpit policy lasted for two years, until the rules were rescinded (presumably because the onus of the rule far outweighed the minute possibility that a pilot or co-pilot would deliberately crash a plane).

The Germanwings crash is the ultimate example of what Scott Shappell, industrial engineering professor, has called a "fad" accident.[11] Fad accidents are so memorable and emotional that they bring on fast changes that solve one problem while perhaps causing others (and not addressing the real issues). Another fad was the call for pilots to be armed with loaded weapons after the 9/11 terrorist attacks in the USA.

Our focus on fad accidents is to be expected; it's a form of cognitive bias called the *availability heuristic*. A heuristic is a "shortcut," meaning that we use this kind of thinking to avoid having to think harder or more carefully – it's almost automatic. When the availability heuristic shows up, it means we only worry about examples that are easy to come up with and, because it's so easy to think of, estimate that it is much more frequent than it is. The classic example is to answer whether there are more words in the English language that start with R or that have R as the third letter.

Take a moment. Which is it?

Did you quickly think of more words that start with R, or more with R as the third letter? Turns out there are *three times* as many words with R as the third letter, but most people think there are more starting with R, because it's easier to think of those. The mental "availability" leads us astray with R-words and also with accidents.

The only way to avoid the influence of fad accidents is to create long-term datasets of disasters, minor accidents, and near accidents. The more boring, but influential, problems will rise to the top and, if these are addressed, will make the biggest impact in avoiding future accidents.

The aviation industry has recognized the importance of gathering data that move beyond one person's unsafe act and distributing the information widely, so everyone can learn from the mistakes of others. Pilots are encouraged to file confidential incident reports whenever they make a mistake, even when they violated procedures, like flying too close to another airplane. Air traffic controllers, flight attendants, mechanics, and ground personnel can also file these reports. Voluntary admission can help protect them from some Federal Aviation Administration (FAA) mandated punishments like suspension of the pilot's license (although they must fit certain conditions, like not having other recent violations and evidence that the incident was not deliberate). In the event of an actual crash, the National Transportation and Safety Board (NTSB) steps in with standard methods to gather data about what led to the event.

Healthcare is just starting to follow such procedures. Physicians have historically held "grand rounds," where they taught each other things they learned and sometimes discussed mistakes they made and how to prevent them. Some of these were video recorded and archived online, but for the most part they were local and had little impact. It is, however, becoming common to publish written case reports detailing errors to reach a broader audience of physicians. It is difficult and humbling for these physicians to admit their mistakes, but such transparency is key for developing systems that avoid similar errors in the future. As the adage goes, "sunshine is the best disinfectant."

Accident Analysis of a Pandemic

> In the midst of chaos, leaders should stop looking for control and start looking for answers.
>
> – Report on the Fukushima Nuclear Disaster by the Initiative for Global Environmental Leadership[12]

The world entered a new level of complexity in early 2020, as an invisible foe spilled across country borders, infecting and killing millions. The novel coronavirus, COVID-19, first appeared near Wuhan, China, a city of 11 million. In just nine months, it killed over one-and-a-half million people worldwide. The United States represented a quarter of the deaths.[13]

A pandemic might seem very different from a nuclear disaster, but because people and their decisions are at the core, they share many similarities. Turning back to the Swiss cheese model, most defenses from the virus are at the *unsafe acts* level of individuals: keep a safe distance from others, wear a mask, isolate if infected, and quarantine if there was contact with an infected person. If these rules were 100 percent observed, the virus could be wiped out.

When I began this section of the chapter I realized that there was no way to include all the ways in which the USA failed to contain the coronavirus. Each layer of cheese had so many holes in its defenses that the number of pages needed threatened to overwhelm the book. What follows is a discussion of one major failure at each level of Reason's Swiss cheese model rather than a comprehensive analysis.

As with Chernobyl, let's move up the ladder in the Swiss cheese model for the "accident" of a global pandemic. Is it true each person could commit or refrain from unsafe acts? What are the variables that affect personal choices in a pandemic? Can the delivery driver refuse to work, or must they continue having contact with the public to feed their family? Does the teacher understand that small children can spread the virus? What options does an essential worker have when their job is indoors in a crowded factory, such as a chicken processing center? In looking beyond the person, we look to the *preconditions* for unsafe acts.

Of a number of possible preconditions, including *complacency, mental fatigue, misplaced motivation, failures to communicate, failures of leadership, and lack of person readiness*, misplaced motivation stood out in the US response.

The motivation should be to follow the guidelines to prevent infection (masks, physical distance, quarantine). But there are other motivations that could thwart adherence: motivation to keep going to a job, motivation to avoid eviction, motivation to prevent foreclosure, and motivation to socially conform to one's political group or gender norms. "Misplaced" doesn't mean that these motivations were not valid or even critical for people in the USA. It simply means that they tended to be in opposition to taking safety measures against the virus.

Above the level of preconditions were *issues with management*. In other areas, management might mean the policies of the nuclear plant or, in aviation, air traffic controllers. For the pandemic, management at the public health level meant local and state government. These were the people who could make policy decisions regarding individual behavior (e.g., a mask mandate) and decisions affecting "misplaced motivation" (e.g., if I choose not to go to work, I will lose my job and my home).

State-level responses varied. Some took decisive action, such as Governor Mike DeWine of Ohio (a state which has almost the exact same number of people as the city of Wuhan, China).[14] Public schools, universities, bars, and restaurants were all closed by March 16, 2020, and a stay-at-home order soon followed. Just as importantly, these actions were backed up with measures to reduce their economic impact. Energy companies could not stop water or electric services. The state removed waiting periods to receive unemployment benefits and allowing temporary childcare centers to start up. The state expanded the list of "good causes" for not accepting work and continuing to collect unemployment (being over the age of 65 was one good cause). Lastly, the state bought back alcohol from the bars and restaurants unable to serve it. These measures addressed the holes in the layers of cheese higher than unsafe acts by allowing more citizens to subsist while staying home. Eight months later, Ohio was performing well in the top third of the states with about 1,400 cases per 100,000 people. Of the eleven states doing better than Ohio, only Pennsylvania had a larger population.[15]

As Ohio patched holes at the management level, Alabama remained focused on individual unsafe acts. In the first four months of the pandemic, Governor Kay Ivey stressed personal responsibility in adhering to hygiene and physical distance guidelines, but with no government intervention. This would be coded under "Issues with management" in Reason's framework. As Alabama soared to the fourth most hard-hit state by the virus, Ivey reconsidered. In a public statement on July 15 of 2020, she announced a statewide mandatory mask mandate by chiding Alabamians with these words:

> You shouldn't have to be ordered to do what is in your own best interests. And in the best interests of those you know and love. Well, folks. I still believe this is going to be a difficult order to enforce. And I always prefer personal responsibility over a government mandate. And yet I also know with all my heart that the numbers, the data, over the past few weeks are definitely trending in the wrong direction.[16]

Appeals for personal responsibility addressed the unsafe acts of individuals, but did nothing for the other layers of cheese. Alabama did not offer government relief to those who lost jobs or to landlords missing rent payments or force energy companies to keep the lights on (though the Alabama companies chose to do this anyway). The only help to the people of Alabama from their state government was a website, ALtogetherAlabama.org, that offered to help individuals "find a job," or "file for unemployment" – both already available pre-pandemic. Personal responsibility can only accomplish so much when the other layers of cheese are forcing people to make decisions about where they go or who they meet, whether it be for a paycheck or childcare. As of early December, 2020, even after the mask mandate, Alabama had the sixth highest positivity rate in the USA (35 percent).

At the final slice of cheese stands the Organizational Influences, which include top-level decisions on budgets, staffing, guidance, and culture. This level, notably the federal government and President of the United States, contained the most impactful failures that cascaded down through the other layers of cheese. In short, the federal government, at the orders of President Donald Trump, misallocated budgets, hobbled staffing, provided incorrect guidance for personal behavior, and created a partisan culture regarding virus protections rather than putting scientists front and center to inform the public.[17] This lack of leadership can be compared to the capable leadership in countries that managed to control the pandemic, such as South Korea.

In South Korea, who had their first case of community spread on the same day as the United States, leadership mobilized multiple systems to stop that spread. Derek Thompson summed up the South Korean capabilities in the May 2020 issue of *The Atlantic*:

> ... some commentators have chalked up the difference to an ancient culture of docile collectivism and Confucianism across the Pacific. This observation isn't just racist. It also exoticizes South Korea's success and makes it seem like the inevitable result of millennia of cultural accretion, rather than something the U.S., or any other country, can learn from right now. The truth is that the Korean government and its citizens did something simple, admirable, and all too rare: They suffered from history, and they learned from it.[18]

Their leaders learned from the MERS crisis in 2015. First, pandemics must be prepared for in advance, including financial investments from the country – a response cannot be cobbled together once a pandemic has begun. South Korea had ready personal protective equipment and ways to create mobile testing sites immediately. Second, politics has no place in keeping the public informed. In South Korea, scientists gave all briefings while politicians stayed silent. This, combined with fast deployment of large numbers of tests, centralized and thorough contact tracing, and effective isolation of those infected, reduced the South Korean numbers to 8 deaths per million people compared to 643 per million in the United States (as of October, 2020).[19]

In an unprecedented publication, editors of the acclaimed *New England Journal of Medicine* analyzed the top-level organizational factor in the US virus spread. In an article titled "Dying in a leadership vacuum," they wrote, "Our current leaders have undercut trust in science and in government, causing damage that will certainly outlast them. Instead of relying on expertise, the administration has turned to uninformed 'opinion leaders' and charlatans who obscure the truth and facilitate the promulgation of outright lies."[20] This cascaded down to the management issues level. Again from the same article: "The federal government has largely abandoned disease control to the states. Governors have varied in their responses, not so much by party as by competence. But whatever their competence, governors do not have the tools that Washington controls. Instead of using those tools, the federal government has undermined them."

And so we can see the Reasons (pun intended) for the unsafe acts by individuals in the United States. Organizational (Federal) influences and lack of planning leading to failures in management at the state level, combined with preconditions across the country, all filter down to the US citizen. Some commit unsafe acts because their jobs or living conditions promote it, some because of scattered information and guidance on how to be safe, and some are unwilling to comply because of misinformation. These human beings are at the core of the pandemic in the same way individuals were at the core of disasters like Chernobyl. But just as with Chernobyl, blaming them without addressing the other layers of needed defenses is not a solution.

Conclusion

Insurance companies often refer to damage caused by earthquakes, tsunamis, hurricanes, and tornados as "Acts of God." It means they consider the damage unavoidable and not the fault of the homeowner. Excepting Chernobyl, it may seem that all of the incidents in this chapter were Acts of God. An earthquake and tsunami bigger than any engineer ever thought possible. A worldwide pandemic, virulent and deadly to some, but with others asymptomatic and unaware of spreading infection. However, even if they were not entirely preventable, they could have been greatly ameliorated. It was our preparation and response to these events that decided their impact. Formal accident analysis tells us where we failed and how to improve. We know it is possible to learn, as South Korea did when MERS struck years ago, but we must make changes for the future and not become complacent when it takes another few decades for disaster to strike again.

7 HISTORY REPEATING?

For a successful technology, reality must take precedence over public relations, for nature cannot be fooled.

– Richard Feynman, Minority Report to the Space Shuttle Challenger Inquiry

The space shuttle *Challenger* broke into pieces seventy-three seconds after launch on a cold day in 1986, killing everyone aboard. Children of the 1980s remember their names: Dick Scobee, Mike Smith, Gregory Jarvis, Judith Resnik, Ronald McNair, Ellison Onizuka, and especially high school teacher Christa McAuliffe. The following days were an endless replay of two diverging smoke columns in the sky, with debris falling down leaving trails like jellyfish tentacles. I still remember the disbelief, the turning to those nearby. Did that just happen? What happened? *How could this happen?*

The engineering reason for the accident was the failure of a rubber seal connecting the fuel boosters. These seals twisted from their position when frozen and allowed flammable gas to ignite. The likelihood and seriousness of the rubber seal issue was known to those at NASA, but even as they tried to redesign the rings they allowed the shuttle to fly. As one of the engineers said in an (anonymous) interview, "I fought like hell to stop that launch. I'm so torn up inside I can hardly talk about it even now."[1] Thus, the *precondition* of faulty equipment was set, awaiting the *unsafe act* of launching in freezing weather. *Organizational factors* and *unsafe supervision* showed themselves when those in charge ignored memos from engineers warning them of the potential seal failures. The contracted company who engineered the seals had known of their issues for almost ten years.[2]

History repeated itself seventeen years later. After launch, foam from the *Columbia* space shuttle broke loose and hit one of the wings, damaging it. The extent of the damage was unknown, and it was decided to keep the possibility of an accident quiet, since the shuttle had to return to Earth. Shockingly, those in charge decided nothing could be done if the damage were bad, and it would be better for the astronauts not to know rather than have them wait in space for their air to run out or try reentry knowing they would not make it. It is unknown whether any measures could have saved the shuttle once it was damaged, though after the disaster many engineers came up with possible solutions that could have been attempted. History, however, is history, and when the space shuttle *Columbia* broke apart upon reentry from orbit, the organizational issues and unsafe supervision at NASA were again exposed. The report of the Columbia Accident Investigation Board stated, "The organizational causes of this accident are rooted in the space shuttle program's history and culture, including the original compromises that were required to gain approval for the shuttle, subsequent years of resource constraints, fluctuating priorities, schedule pressures, mis-characterization of the shuttle as operational rather than developmental, and lack of agreed national vision for human space flight."[3]

The causes of the accident weren't only the engineering failures. The causes were rooted in a history of high pressure launches with a get-it-done culture that wasn't funded well enough to perform. The report goes on to say, "the NASA organizational culture had as much to do with this accident as the foam."

Why do we repeat the same mistakes again and again?

The 2010 explosion and oil spill in the Gulf of Mexico from BP's Deepwater Horizon drilling rig killed eleven people immediately and released five *million* gallons of oil into the waters, tainting and killing sea life for years after. The spill followed the same patterns as the previous tragedies with faults in multiple systems lining up for a major accident.[4]

The story of the oil spill began three miles under the waters of the Gulf of Mexico. The oil company planned to cap a drilling site before moving on to the next. Workers poured cement to seal the site, but the seal allowed gas to build up. The gas caught fire in an explosion that destroyed the rig. The committee report from the National Academy of Engineering called out the *preconditions* of poor engineering of the concrete and some other systems, but also the *unsafe supervision* of those deciding whether the concrete cap was sufficient. The backup systems designed to cut off flow in the event of an emergency were not designed for the situation and failed. In terms of *organizational factors*, there were multiple contractors and teams working on different portions of the project, and little standardized communication between them. The *unsafe act* directly related to the explosion was the acceptance of ambiguous test results regarding pressure by the workers in charge. There was no requirement to check those results with experts (*unsafe supervision*).[5]

Eerily similar to Chernobyl, the committee's report called out that "Overall, neither the companies involved nor the regulatory community has made effective use of real-time data analysis, information on precursor incidents or near misses, or lessons learned in the Gulf of Mexico and worldwide to adjust practices and standards appropriately."[6] The report called out the culture of secrecy in oil, saying that until there is less litigation and more public demand for transparency, the industry will refuse to share the information on accidents and near misses to create a culture of learning, safety, and improvement. This is a superficial treatment of the numerous factors in the disaster, but shows how the Swiss cheese model can organize and categorize factors for improvement in many industries. Reason's model explains the accidents even as we seemed doomed to repeat them.

Surely we know by now how to build better systems? In many instances, we do. Air travel is many times safer now than it was forty years ago.[7] Oil spills have declined precipitously since 1970, despite more oil being moved and used.[8] We still fall short, but looking at every slice of cheese (often called "systems thinking") in our most complex systems has made them safer. However, it is all too human to repeat the sins of earlier generations. We tend to become complacent when systems have been working well, and the better they are working the less we perceive that we need them. It is harder to recognize threats when the systems we have created have protected us from them, leading us at times to purposefully dismantle those systems and at times to inadvertently dismantle them or dismantle them through neglect. Improvement is only the first step – vigilance is still required.

Attend to the Mote in Thyne Own Eye

> No man is an island,
> Entire of itself,
> Every man is a piece of the continent,
> A part of the main.
> If a clod be washed away by the sea,
> Europe is the less.
> As well as if a promontory were.
> As well as if a manor of thy friend's
> Or of thine own were:
> Any man's death diminishes me,
> Because I am involved in mankind,
> And therefore never send to know for whom the bell tolls;
> It tolls for thee.
>
> – John Donne, 1624, meditation no. 17 from *Devotions upon emergent occasions*

In systems as complex as space exploration, nuclear power, aviation, healthcare, and a multitude of others, the Swiss cheese model shows the importance of design at all levels, from the controls used by a worker to the culture of the organization as a whole. A good system keeps us safe and controls our worst natures: laziness, complacency, bias, and risk-taking.

One of the most famous (or notorious) psychology studies of the twentieth century showed just how important a good system was, even outside of the machines, automation, and other trappings of industry. Philip Zimbardo, a professor at Stanford, was interested in how the environment affects behavior and identity. He randomly chose young men to play the part of prisoners or guards and had them live in the basement of the psychology building on campus in their assigned role. The guards, like real guards, went home each night. The prisoners stayed in the "prison," even though there were no real locks keeping them inside.[9]

It wasn't long before the guards started to mistreat the prisoners: required pushups with a guard's foot upon the prisoner's back; stripping them naked as punishment; restricting toilet use and forcing them to defecate in buckets. Each person fulfilled their "role," including Zimbardo himself as prison warden rather than social psychology professor. The guards weren't given much training and only a few rules, which some followed and others did not. It was not until an outside observer called them out on the immorality of the study that it was ended. Essentially, when a human system was left to self-organize and regulate, conditions became inhumane.

Zimbardo concluded the solution was transparent and firm oversight. People outside of the situation need to know what is going on and set rules ahead of time with frequent checks. The individual guard, or prisoner, or warden, is just one piece of a larger system that needed as many safety precautions and inspections as a nuclear reactor.

I wish I could say that we learned the lessons of the 1970s Stanford prison experiment, but more recent history shows we have not.[10]

In 2003, the United States was two years into war with Iraq, based on the erroneous premise that Iraq had weapons of mass destruction. In Abu Ghraib, an Iraqi prison was transformed into a US-operated detention center that held thousands of Iraqi prisoners. The prison was run both by the US military and a contractor, CACI International. News of abuse of prisoners trickled out, but it wasn't until April of 2004 that *60 Minutes II* broadcast some of the graphic photos showing members of the military violating their own protocols and the international conventions against torture.[11]

The first mainstream news coverage of Abu Ghraib was disturbing: naked prisoners on leashes held by US personnel, hooded prisoners convinced they were going to be electrocuted, naked piles of bodies with their captors giving the thumbs-up sign from behind. The photos released to the public weren't even the worst ones – years later, when the administration changed to President Barak Obama, the president prevented the release of the full set of photos, stating how damaging he believed they would be for the United States' reputation and citizens worldwide.

The mistreatment of prisoners at Abu Ghraib by US troops was an eerie twin to the Stanford prison experiment and was due to the same failings of oversight. The public blamed the soldiers for their actions, and their actions deserved blame. But such actions were only possible in a system with *unsafe supervision*, poor *preconditions*, and an *organizational culture* that saw the enemy as less than human. As with Chernobyl, Fukushima, the *Challenger*, the Deepwater Horizon spill, and the Stanford prison experiment, we can march upward through the slices of cheese identifying the holes in each.

At first, the US government was quick to lay blame on the "bad apples" at Abu Ghraib rather than take responsibility. In an early interview on *60 Minutes II* with Dan Rather, General Mark Kimmitt said, "I think all of us are disappointed by the actions of a few. Every day, we love our soldiers, but frankly, some days we are not always proud of our soldiers."[12] Later he acknowledged the failures of leadership, but at first there was focus on the soldiers who carried out the mental, physical, and sexual torture of the prisoners. The commander of the battalion that included the soldiers who abused the prisoners was asked in an interview, "Who would you pin the responsibility on the actions of those individuals at Abu Ghraib?" And she responded, "The [Military Police] that were involved. That's who I'd pin it on and I'd pin it on [redacted] the Platoon Sergeant, and the First Sergeant, Captain [redacted]."[13]

All subsequent reports identified serious issues above the level of any individual soldier. Major General Antonio Taguba listed so many supervisory issues at Abu Ghraib in his 2004 report that they may only be summarized. Lack of adequate staff and resources, overcrowding of prisoners, poor training in how to be an occupying force, low morale, and higher than normal attacks on the prison were the physical, social, and psychological preconditions for a major incident. One note that stood out was that the soldiers had been led to believe they would be going back to the USA at the end of a previous mission. Instead, they were assigned to Abu Ghraib, trapped in a prison under fire with no dining hall, barbershop, stores, or recreation facilities.[14]

Supervision at Abu Ghraib tacitly supported the actions of the individual soldiers and it was well known that even if the soldiers were disciplined, it would be light. The Taguba report noted the official journals were sometimes flippant, with notes like, "Strollthrough; All Secure, all terrorists present." The unsaid message sent from supervisors was that they weren't really watching and didn't care what happened. The Taguba report stated that because of conflicts in the command structure and no clear guidance about responsibility, "Coordination occurred at the lowest possible levels with little oversight by commanders." This lowest possible level was the level of those directly in charge of the prisoners – just like at Stanford.

These lower-level indiscretions would be stamped out quickly if there were no holes in the upper levels of cheese. Unfortunately, the tone of interaction with detainees and prisoners at Abu Ghraib was set from the top. Though the USA was part of the Geneva Convention against torture, President Bush signed a memo in 2002 that stated, "none of the provisions of Geneva apply to our conflict with al Qaeda in Afghanistan or elsewhere through the world."[15] In 2008 the US Senate released a report of their own investigation into the treatment of detainees in American hands (not just at Abu Ghraib) with this damning forward:

> The abuse of detainees in US custody cannot simply be attributed to the actions of "a few bad apples" acting on their own. The fact is that senior officials in the United States government solicited information on how to use aggressive techniques, redefined the law to create the appearance of their legality, and authorized their use against detainees. Those efforts damaged our ability to collect accurate intelligence that could save lives, strengthened the hand of our enemies, and compromised our moral authority.[16]

The committee included many senators, notably John McCain of Arizona, who had himself been tortured decades before in North Vietnam. In an interview with *Face the Nation* in 2014 he said,

> What we need to do is come clean, we move forward, and we vow never to do it again. That's what we did after Abu Ghraib, and that's what we've done after other times in our history. We're not a perfect nation, but we are a nation that acknowledges our mistakes and we move forward. And we are not going to be inhumane.[17]

These were heartfelt words, and ones he clearly believed. However, the fact that he was being interviewed a dozen years after Abu Ghraib about the US CIA using torture such as waterboarding meant the system hadn't sufficiently improved.

I hesitated to include the Stanford prison experiment as an example in this chapter, because it was an experiment that should never have been done and would not be allowed today. Yet we learned from it the human tendency for an unmonitored and non-transparent system to devolve and need apply that lesson in our future regulation of human systems. To make the experiment "worth it" we need to apply those lessons more thoroughly. Human factors experts Scott Shappell and Douglas Wiegmann created a formal way to do exactly that: The Human Factors Accident Classification System (HFAC).[18]

Acknowledge Mistakes and Move Forward

The HFAC system, based on the Swiss cheese model, is a guided checklist for what to look for after an accident. The first step in an HFAC analysis is to understand the potential for failures at all of the levels in the Swiss cheese, especially the hidden ones at levels above the actions of one person. The checklist can be used to analyze accidents that already happened or to assess the strength of a system to prevent an accident from occurring. This includes looking at the potential for unsafe acts by individuals, but also a check on the potential preconditions, the functions of management, and the culture of the organization. It's easy to detect issues in hindsight, but a good checklist gives foresight.

Those analyzing the unsafe acts are prompted to consider whether the person made an error, which would be something inadvertent, or a violation, where there was a well-known rule and the person decided not to follow it. Mis-assessing the pressure readings in the Deepwater Horizon spill was an error because the workers didn't have the knowledge or training. But continuing to fly shuttle missions while knowing the risk of the frozen rubber seals was a violation. Keeping informal notes on prisoner activity at Abu Ghraib may have been an error, but beating prisoners was a violation. Daily life has the same division. Turning right while the stoplight is red is normal in most cities, but illegal in New York City. Driving in that city, it's likely one might make that error. However, a New Yorker who turns right on red knows what they are doing – making it a violation.

This distinction is important because you fix errors differently than violations. "Errors" mean people need more training or systems to prevent the error. "Violations" mean there is a problem in the organizational culture where people feel the need to go off on their own rather than follow the rules, or that the system is so restrictive that people seek easier (but unsafe) paths to complete their jobs. You can go even deeper here, by looking at errors from lack of skill, errors that are from bad decisions, and perceptual errors. Each one has a different "fix." The law may treat them the same – the NYC visitor will get the same ticket as a native – but one ticket serves to educate and punish while the other only punishes. The poor non-New Yorker would have much preferred a system that made it obvious right on red was not allowed instead of getting a ticket.

It Tolls for Thee

Humans are social creatures – so social, in fact, that solitary confinement of more than fifteen days is considered torture by the United Nations.[19] In most cases our social natures promote survival. Richard Wrangham, a professor in the Department of Human Evolutionary Biology at Harvard, noted the difference between us and our primate cousins in a 2015 interview: "If you put twenty chimps on a jet plane and tried to send them across the Atlantic, let me tell you that only one or two would walk off that plane alive. We do this all the time. We take it for granted as human beings that big groups of people can get along with one another."[20] We may have our human squabbles during air travel, but being murdered by the rest of the passengers isn't one of them. It's telling that we worry more about other passengers judging us than we do about violence. Being able to cooperate in groups is the reason we have complex systems like airplanes in the first place.

However, as with every all-too-human attribute, there is a downside to our social nature. Social pressure, both intentional and unintentional, can cause errors and violations. Susannah Paletz, at the time a research psychologist at NASA, sought to describe and include these pressures in the HFAC checklist.[21] Along with her colleagues, she collected the types of social pressures that caused pilots to take unnecessary risks. When pilots observed others succeed, especially those they admired, they felt pressured to try the same. One pilot said, "He's got his reputation to live up to as far as, well, three other pilots made it; what's wrong with you?" This is an error if their own skills were not up to the task or if conditions changed. On the violation side, she included "normalization of deviance" by supervisors, when pilots knew they shouldn't fly but felt encouraged to do so. One pilot was quoted as saying, "[A manager would say] 'Why don't you go take a look? See what it looks like. It's legal to leave – go look.' You get out there and generally you don't come back. You're already out there." The supervisor didn't command the pilot to fly, just nudged them in that direction with a little social pressure.

Those same social influences were present at Abu Ghraib. Normalization of deviance was probably the most prevalent, given the reports of supervisors "broadening" the accepted interrogation methods. Techniques far below what actually occurred (such as not allowing a prisoner dinner) were considered serious enough to require approval by the commanding officer. But that changed during the war as the commanding officer gave blanket approval. Those techniques became normal and new techniques were invented by the soldiers doing interrogation. In 2004, the *Baltimore Sun* published interviews with some of these soldiers who listed the social pressures, such as, "There would be the handoff from MI [Military Intelligence] to the MPs [Military Police], and the word would be, 'Here you go, here's one who's not cooperating,'" one of the soldiers said. "Then – What do you know? – that prisoner ends up beaten or paraded around naked." When one soldier attempted to report these violations, the social norms were further cemented. "I was told, 'Don't worry about it – they probably deserved it.'"[22]

Again, these social influences are part of being human. We are the ones who have to put safeguards and enforced regulations in place that prevent our social tendencies from becoming errors and violations. If left to ourselves and pressured to perform in extreme circumstances, we may end up looking more like the chimpanzees on a plane than an advanced and civilized society.

Moving up to analyzing preconditions, the HFAC system asks about the mental and physical states of the workers: Were they tired? Bored? Angry? Focused on the wrong problem? Medicated? What was the weather like? How did the outside environment affect behavior? We covered some of this in the first discussion about Abu Ghraib – conditions were stressful and the soldiers stationed there had poor quality of life compared to other assignments. As one soldier said, "We were working 12-hour days, sometimes more, six days a week – and then catching up on the seventh day."[23] The commander of the battalion, Brigadier General Janis Karpinski, was not even allowed to travel to or inspect the prison after dark. In a 2005 interview for PBS *Frontline* she said, "We could not travel after dark, so I was not allowed to go out to Abu Ghraib or any other facility after dark." And in reference to the damning photographs, "But it was allegedly or apparently by design, a specific night, specific hours, when these kind of activities would be conducted, because there was no chance at all of Gen. Karpinski dropping in, because they knew that there were restrictions placed on travel after the hours of darkness."[24]

To underline how important the upper layers of cheese are, I'd like to point out that many of the stressors on guards at Abu Ghraib were not present in the Stanford prison experiment. The lives of the student "guards" were only altered when in the psychology building on a college campus. Yet even those students, going home each night to their normal lives, still mistreated prisoners due to their "expected" roles. What could we expect from the Abu Ghraib soldiers?

When considering supervision, there are four areas the HFAC system considers: First, the competence of managers and supervisors. Are they making sure everyone has the right training? Are they watching what goes on and offering feedback? Do they prioritize safety? Second, do the managers ask workers to do things that are dangerous? Do they ask them to work too quickly? Third, when a supervisor is aware of a problem, do they work to fix it? This might be by reporting it or correcting it themselves. Lastly, do they encourage or discourage following the rules? This layer at Abu Ghraib seemed to have so many holes it barely existed. An anonymous soldier said of the commander, Colonel Thomas Pappas, "We never saw him" and "He ate, worked and slept in one room. So it's like nobody's in charge."[25] Pappas himself was under high pressure to get and pass along useful information from prisoners, which likely also contributed to unusual decisions like allowing soldiers to bring in intimidating dogs to the interrogation rooms.

Finally, examine the organization cheese slice. Did the organization provide enough resources to realistically maintain safety? What was the culture and climate? Did workers feel appreciated? Pressured to perform faster or succeed "at any cost"? What rules and regulations did the organization insist upon? Did they make these processes well known, required, and easy to follow? Was there a sense that it was "better to ask forgiveness than permission"?

Brigadier General Janis Karpinski, in charge of the military police at Abu Ghraib, was vocal about the attitudes high-ranking officials brought to the prison. On PBS *Frontline* in 2005 she recounted when a new leader was introduced and told her, "You have to treat the prisoners like dogs. If you treat them, or if they believe that they're any different than dogs, you have effectively lost control of your interrogation from the very start."[26] In her official interview with Major General Taguba she also mentioned another general's attitude toward detainee abuse: "General Hill kind of minimized it. Not kind of. He minimized it and he said 'These things happen.'"[27] Karpinski herself did not leave unscathed, as Taguba included in his official report, "What I found particularly disturbing in her testimony was her complete

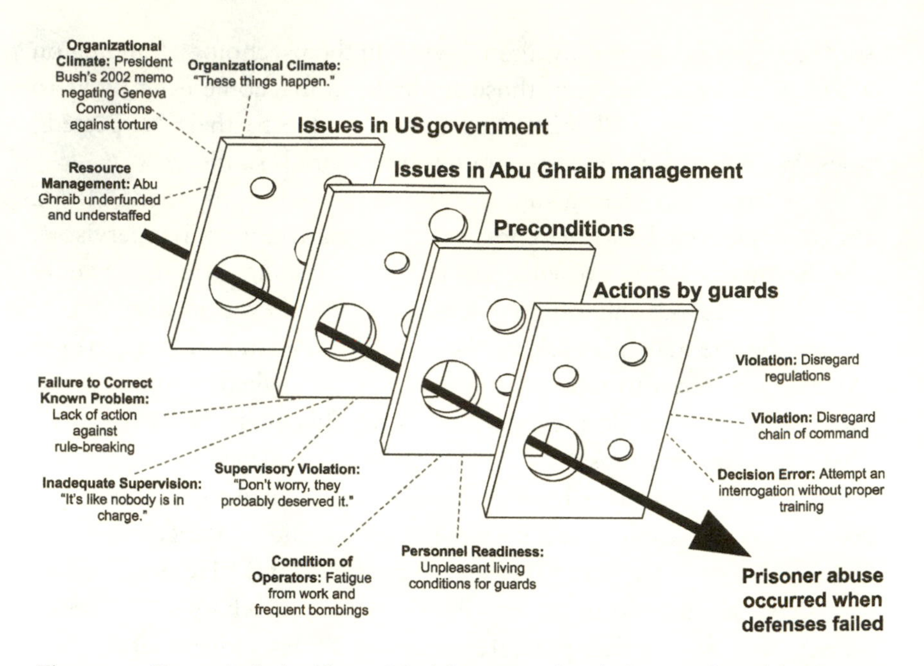

Figure 7.1 Reason's Swiss Cheese Model annotated with the specifics of the Human Factors Accident Classification system and details from the prisoner abuse scandal at Abu Ghraib detention center in Iraq, 2004.

unwillingness to either understand or accept that many of the problems inherent in the 800th MP Brigade were caused or exacerbated by poor leadership and the refusal of her command to both establish and enforce basic standards and principles among its soldiers."

In the HFAC system, each layer of the Swiss cheese gets specific treatment, with each item relating to how to fix that item (Figure 7.1). Using it to analyze multiple accidents should find the most common holes in the Swiss cheese of an industry and offer solutions. Each potential error or violation can be mapped upward to solutions at higher levels to help prevent them in the first place.

Conclusion

It's never just one thing. It's rarely just one person. Blame, especially blaming the victim, does not prevent future accidents. Only a careful look at all the contributing factors and their fixes can do that. One of our most amazing achievements as humans has been the systems we create to overcome the limits of our minds and bodies.

Sometimes these complex systems, from space shuttles to oil rigs, fail us disastrously. Thus, we cannot forget to use the mental support tools we have created. These tools, such as accident analysis (e.g., HFAC), formalize our thoughts and ensure we don't miss subtleties or hidden dangers in our complex systems. The saying "safety is no accident" is trite but true. It takes planning, culture, oversight, regulations, inspections, well-trained supervisors, and workers who are supported as much as they are depended upon. But we must take our lessons learned into the future and not allow complacency to infiltrate our thinking. History is only doomed to repeat itself if we allow it to do so.

8 NEEDLES IN HAYSTACKS

A shepherd-boy wishing to amuse himself at the expense of his fellow-villagers, came one day running along, crying "Wolf, wolf!" as if one of those ravenous animals had attacked his flock. The people, eager to defend the sheep, bestirred themselves; but when they came to the place, they found no wolf there. So, after scolding the young shepherd, they returned home. A few days after, a wolf really did fall upon the boy's flock, whereupon he ran away to the village crying "Wolf, wolf!" with all his might. The people told him they were not to be imposed upon twice, and they therefore were resolved to pay no attention to his cries.

– W. and R. Chambers, 1839, *The moral class-book*

"I don't know," Abby said, "It was kind of awkward. He was kind of awkward."

"What happened?," I asked.

"Too many weird pauses. I mean, on paper he seemed great. Totally into hiking, likes the same TV shows I do, but in person he just doesn't seem to have anything to say to me." She lifted her eyebrows and looked up at the sky. "Awkward."

My friend Abby was newly single and on several dating sites. "You see the same guys on all the sites, though," she noted. Many of the sites proposed matches, using unknown algorithms, and Abby was usually game to go on a date with a good match. "I've met some nice guys," she admitted, "but nobody I really clicked with yet."

The challenge for Abby, and for the dating sites themselves, is to separate a signal from the noise. The signal is, in Abby's case, a match with a life partner. The noise is all other candidates: a quick search in our geographical location turned up over 2,000 potential mates.

Abby started separating signal from noise by searching only for those with her preferred qualities: men up to five years younger or ten years older, a particular education level, and other beliefs and behaviors she (thinks) she desires. Even then, she turns up over 1,300 potential dates within a 20-mile radius. Here's where the dating site tried to add value: she also filled out extensive surveys about herself. The site's algorithms supposedly analyzed those answers and matched her to a smaller set of results.

For Abby, there are four possible outcomes from the dating site. What she wants most is a "hit." A hit is when the site suggests a guy for her and she agrees – best case they fall in love, get a dog, and live happily ever after. Unfortunately, what she's been finding most frequently is a "false alarm." A false alarm occurs when the site suggests a guy but she doesn't end up liking him. There are two other possibilities: a "miss," who are guys that would be great for Abby, but the site never suggests them, or a "correct rejection," who are the guys that are a poor match for Abby *and* the site recognizes that and doesn't suggest them to her. These four outcomes are the classic outcomes of *signal detection theory*, which applies to terrorist attacks, cancer diagnoses, and nuclear destruction just as well as it applies to Abby's dating life.

Because Abby is almost always willing to go out, even if the fellow appears to be a less-than-perfect match, her strategy leans toward saying "yes" to a date. But my friend Seth has the opposite dating strategy; he only goes on dates with the tiniest subset of suggested matches. It's a rare weekend when I hear he's going out with someone. Saying "yes" to everyone means that Abby is going to have more false alarms than Seth, but she also might end up with more hits. Indeed, other friends give the same advice to Seth again and again – "Just go out. You'll never know if you don't give it a chance." Those friends are trying to push his strategy to be more liberal. (Liberal and conservative in signal detection theory have no connection to political parties, thank goodness!)

Abby and Seth are both drawing a line – any signal stronger than that line means they agree to a date but any weaker means they pass on that person. Obviously, Abby and Seth draw their lines in very different places on the dating spectrum. But both of them subconsciously know that there is some continuum of an unknown quality they are using to decide. For Abby, it's "attractiveness to Abby." For Seth it's "attractiveness to Seth." Abby is willing to take a chance on a guy she thinks might be attractive, but Seth isn't willing to go out unless he's pretty sure his date is right for him. Signal detection gives these different decision criteria a name: it's their response bias.

For an intelligence analyst for the military, that unknown quality they are judging is "resemblance to the enemy." For a radiologist it's "resemblance to a tumor." All these continuums go from low to high. At some point Abby has a threshold of attractiveness where she starts saying yes to dates with guys above that threshold. Unfortunately, there are probably guys above that threshold who are *not* a good match for her, but she will have to go on a date with them to be sure. And there are probably guys below her threshold that would be a great match for her, but she'll never meet them. There is a sweet spot where the probability of someone being a match is equal to the probability of them not being a match – this is the threshold where she can maximize her hits while minimizing her false alarms. But, since we often don't choose this ideal point, we (and Abby) are biased.

Signal Detection Theory as a Way of Thinking

Hits, false alarms, misses, and correct rejections are the core of *signal detection theory*.[1] The reason it is called a theory is that once you divide results into those four types, it explains a good deal of human decisions and behavior. Dividing the world up into those four categories is addictive – nearly every news story has me begging for more information to fill out the grid in Figure 8.1.

The grid in Figure 8.1 shows a representation of signal detection that can apply in any situation where the best decision isn't known. In Abby's case, she is trying to guess whether someone is a match for her. This is a vague trait, since it's a balance of personality, looks, interests, and values. Of course, every man she goes on a date with also has his own signal detection going on, and also has to find her a "hit" at the same time Abby thinks he's a "hit." You can see why love is so hard to find.

		Signal Present	
		Yes	No
Response	Yes	Hit	False Alarm
	No	Miss	Correct Rejection

Figure 8.1 The four possible outcomes of a decision in signal detection theory.

The beauty of signal detection theory is that it can quantify the effectiveness of any system looking for signals hidden in uncertain situations. It doesn't matter whether that system is a human or a machine, the only three pieces of information needed to judge effectiveness are response bias, the difference between the signal and the noise, and the base rate of signal occurrence.

The High Cost of False Alarms

The most (almost literally) earth-shattering example of the high cost of false alarms was the Cuban Missile Crisis in 1962. The USA almost launched planes armed with nuclear weapons, partially due to false alarms. The last trigger was a supposed saboteur discovered trying to enter a military compound. President Kennedy had already set the level of alert to DEFCON 3 (DEFCON 5 indicated peacetime, DEFCON 1 indicated war) based on Soviet missile construction in Cuba. Pilots knew that there were no drills planned when at DEFCON 3, so when the alarm sounded, they were certain World War III had begun. But it was a false alarm – the saboteur turned out to be a curious bear climbing a fence, fortunately identified before we mistakenly started a nuclear war with Soviet Russia.[2]

If that weren't a close enough call, just two days later a US ship dropped warning charges near a Soviet submarine off the coast of Cuba. The captain of the sub interpreted the warning as an attack and ordered their nuclear missile readied. The false-alarm detecting hero in this case was the 2nd in command, Commodore Vasili Arkhipov, who refused to authorize the start of World War III without explicit orders from headquarters and would not fire on the US ship.[3] Indeed, the Soviets deserve credit for multiple instances of cool-headedness under stressful false alarms. In 1983, Soviet forces received word that the USA fired five nuclear missiles at the Soviet Union. All systems reported the missiles as real and on their way, but it was due to a satellite misinterpreting sunlight on clouds as a missile launch. Fortunately, Col. Stanislav Petrov was in charge and didn't completely trust the satellite system (and had a conservative response bias when it came to starting a nuclear war). He correctly classified the incoming missiles as a false alarm instead of retaliating.[4]

We do not know the "correct rejection" rate of these threat detection systems, but we do know every alert thus far was a false alarm. We also know that the hit and miss rates are (thankfully) both zero, otherwise I would not be here to write this chapter.

If You Think Love Is Hard to Find, Try Finding Explosives

The fate of the TSA (and occasionally, airplanes) depends on signal detection. Here, the signal is any disallowed object on a plane: a weapon, a large container of liquid, or flammable materials such as camping fuel. The noise is every other conceivable object a human can put into a bag. Patrick Smith, who ran an "Ask the Pilot" column, said, "Of course guards are going to fail when you've assigned them an impossible and unsustainable task: the detection and confiscation of every conceivable weapon, from screwdrivers to automatic firearms, from more than two million air travelers every single day of the week."[5]

When the TSA takes your bag for further screening, but then gives it back to you, you know they just had a false alarm. When no notice is given to your bag at all, they've "correctly rejected" the idea that you're carrying a weapon. Unless, of course, you are Blake Alford, who accidentally packed a loaded, semi-automatic pistol in his carry-on in Atlanta and only realized it when he got to Chicago. In an interview with CBS, Alford referenced false alarms compared to misses: "They'll make people get out of wheelchairs. They'll make me take off my belt buckle and my shoes. How did my gun go through?"[6] For the TSA in the Atlanta airport, this was an embarrassing "miss."

Unfortunately, there is no easy cure for stopping misses and false alarms. But decades of research in psychology has revealed some helpful tricks and tips. These include changing someone's response bias, improving their sensitivity, and using technology to better separate the signal from the noise. This can be done through training, experience, partially automating the detection with computers, and altering the user interface that aids the TSA agent in detection.

The Training Option

The rarity of weapons and illegal objects in luggage is what makes the job of the screener so difficult. If only one bag in 500 has a knife and one bag in millions has a bomb, it is easy to get used to glancing at a screen and responding that you see nothing. This is a little counterintuitive, since you might think a knife or bomb would stand out because of its rarity, but we have evolved to see what we expect to see. Usually, this is a feature rather than a bug – there wouldn't be much reason to always be on the alert for tigers if the last tiger existed only in the tales of your grandparents.

Jeremy Wolfe, director of the Visual Attention Lab at Harvard, led an experiment where the rarity of finding a weapon was changed for TSA agents.[7] His hypothesis was that increasing the chances of a weapon showing up, even artificially, would result in the agents getting more hits and fewer misses. This is a phenomenon well known in basic research on visual search called the *prevalence effect*. But Wolfe had another idea as well: this enhanced detection would last even after the target rate returned to normal. He predicted that even when the targets went back to their "rare" status, the agents would be still affected by the earlier hits and thus more likely to keep up their high-hit, low-miss performance.

Real, newly trained, TSA agents were in the study. They first practiced with x-rays that had very few illegal objects (less than 5 percent). As expected, they missed quite a lot, got some hits, and didn't have many false alarms. Then, Wolfe hit them with a stretch of bags where *half* of the bags had an illegal object. The agent's hit rate shot up, and they hardly missed any during this section of the experiment. Then, Wolfe changed the rate back down to less than 5 percent. The agents maintained their high hit-rate. This meant having more false alarms (a more liberal response bias), but when it came to airport security, a *few* more false alarms were better than missing contraband.

Companies have already applied these results in software products for the TSA. Rapiscan, the company that makes many airport scanners, includes a software option called TIP (threat image projection). They advertised, "At configurable frequencies defined by a supervisor, TIP inserts digital fictional threat images such as guns, knives and bombs as if the threat object were actually packed inside the passenger's bag into the regular flow of bags displayed on the x-ray system monitor."[8] It's a real bag the agent sees, but the image of contraband has been seamlessly inserted in the x-ray photo. Inserting known fictional threats allows extra feedback to the scanners and reports on their performance to their supervisors – they know when the scanner should have caught something going through. Though including fake threats to identify has some benefit for the workers (almost nothing is more draining than searching for targets that never appear), the constant supervision affects morale and the decisions the agents make about bag searches. Agents have reported that they felt constantly watched but rarely rewarded, only punished if they made a mistake.[9]

Barbara Peterson was a journalist who went undercover to become a TSA agent. She wrote a piece for *Condé Nast Traveler* that described her training, the attitudes of the agents and administration, and her experience with TIP. She said,

> I am not permitted to monitor the machine unattended, even though I had hours of practice in class. When the line slows, I get a crack at it with a seasoned screener at my side and quickly learn how to spot a lighter, the most common of the prohibited items we see (at least a dozen are collected by the end of each shift). We're also on our toes to spot images of dangerous items that the TSA briefly superimposes onto the contents of bags to catch us if we let down our guard.[10]

It was clear to her that these fake threats weren't there to improve her searches, only to trip her up. Another downside of adding fictional threats to the x-rays is that screeners will get used to whatever system provides the punishments and payoffs, rather than focusing on actually detecting weapons. Doug Harris, a human factors psychologist, detailed the story of one TSA agent.[11] An agent, familiar with the TIP system, was presented with a fake bomb physically placed in a bag. The screener saw the bomb, and pressed the key to indicate he saw something. But the TIP system didn't give him feedback saying he caught a weapon, because the fake bomb was a physical object, not a fictional overlay provided virtually through TIP. He was so used to all threats coming through the TIP system that the lack of feedback made him assume he saw it wrong and went back to his job without flagging the bag for further screening. Harris quotes the agent as saying, "Alright, that can happen, it just looked like a bomb, next time I will press the button only if I am really sure there is a TIP." That agent learned exactly what he was taught through experience: threats come from the TIP system and are fictional. But what we want the screeners to learn is how to keep bombs off of airplanes.

Automation to the Rescue?

TSA agents no longer have to depend on their eyes to parse the thousands of greys in x-ray images: automation color-codes the type of material (metal, plastic, wood). Some go so far as to highlight items the automated system believes suspect, but leaves the final call to the agent. The newest scanners map the shape of all 3D objects in the bag, and the TSA agent can electronically rotate the objects to see them better. Some of these improvements are increasing the strength of the signal, helping agents to get hits and fewer false alarms. But in some cases the increased automation is a beautiful microcosm of what human factors psychologists call the "paradox of automation."[12] Making the TSA system more automated might seem like a great idea, but, as with every great idea, "it's complicated."

Levels of Automation

Automation at its lowest level is called "information automation." Color-coding materials fall into this category – it gives the person extra information to make the decision, but doesn't suggest a course of action. Information automation is everywhere: graphs of historical stock performance, a low-gas light in a car, and graphics showing the remaining seats on a plane, just to name a few.

Once the system starts highlighting suspicious items though, it rises to the level of "decision automation." It's not just giving information, it's suggesting an item is a threat (even if the agent makes the final decision). An even higher level of automation would be a system that identified the threat *and* made the call itself, sending a bag for more invasive screening, and only stopping if told to by the agent.

The highest level of automation is when no human decisions are allowed and the machines run on their own, usually because we are slow and imperfect compared to computers. Missy Cummings, former fighter pilot and current Human Automation Interaction (HAI) researcher, described that level of automation in being catapulted from aircraft carriers in the 1980s: "I'd have to turn to the left, show my hands, that I wasn't touching anything, then turn to the right, show my hands that I wasn't touching anything, then turn to the front and grab two bars, where everyone can see your hands. And only then, when they were sure the pilot couldn't mess anything up, would they launch the catapult."[13] She noted that experiences like that got her interested in the human factors of automated systems.

When it's working well, all of these automation methods result in increased hits and the best balance of misses and false alarms. Since the screeners most want to avoid a miss, this is a good thing! But we can end up on the dark side. The paradox of automation is that as much as it helps humans when it works well, it can lull us into complacency. We mindlessly follow suggestions, get bored, and become likely to miss the times the automation fails. And all automated systems will fail at some point. The two attributes most likely to encourage complacency are (1) a system that almost always works perfectly and (2) when most of the signal detection is automated, rather than forcing a person to consider all the evidence.

Changing the Payoffs

How do we help people adjust their judgments? How can we encourage them not to be complacent, or too trusting? Or to encourage a wary user to take better advantage of the automation and stop trying to do everything themselves? As mentioned earlier, one way is through experience and feedback. Hand in hand with feedback is an adjustment of the payoffs for a decision. A payoff is just a reward or punishment. A payoff might be a bonus for finding more weapons in the TIP images *or* it might be getting fired for missing a TIP image of a bomb. Dating sites would probably work a lot harder on their matching algorithms if they got a million dollars for every marriage that came from their site. Conversely, Abby wouldn't be so keen on giving new guys a chance if a few of her dates veered past "awkward" and more towards fiasco.

Rewards and punishments affect the strategies people use in every uncertain decision, from personal interactions to policy, to legal battles. Imagine being a radiologist. Patients are usually happy to be told they don't have cancer. They are also relieved to discover they don't have cancer after thinking they might have cancer. What people do not like is being told they do not have cancer, but then finding out the radiologist missed it, and that now it's too late for treatment. Aside from anger and hurt, this also results in lawsuits. Radiologists are human – they diagnose using both their knowledge of the likelihood of a kind of cancer and a response bias toward favoring false alarms over misses. They are biased toward telling a patient that they might have cancer and ordering further tests to confirm, even though many people end up not having cancer after all. The punishments for a miss push them into a liberal response bias (and causes more than a few unnecessary biopsies).[14]

Better Data, Better Signal Detection

> ... repeated exonerations of convicts under death sentences, in numbers never imagined before the development of DNA tests.
>
> – *Kansas v. Marsh*, 2006, Supreme Court Justice David H. Souter

As of October 2020, 375 prisoners in the USA had been exonerated using DNA evidence.[15] One hundred thirty of these were erroneously convicted of capital crimes, many facing a death sentence. Before their innocence was proven, they served an average of fourteen years in prison; without DNA analysis most would still be there. DNA analysis is an important tool in revealing more of the true state of signal detection in criminal justice by showing a better picture of the ratio of hits and false alarms. This illuminates (and will perhaps change) our response bias for conviction. Before DNA analysis, the US public was confident these convictions were just and the right people were in prison.

Court systems have always been biased toward conviction, and though false alarms were not desired, they were also mostly unknown. The amount of false alarms revealed through DNA analysis showed a more accurate picture of the justice system bias. Though there are other variables, the advent and more widespread use of DNA testing in the mid 1990s and subsequent exonerations has paralleled a steady decline of approval for the death penalty by the public as well as a change in the response bias of our court systems.[16]

DNA analysis provided the incontrovertible evidence of a false alarm. The most interesting side-effect came from how it pointed out what evidence in each case wrongly led to conviction. In 75 percent of the cases where DNA evidence showed the convicted *could not have been the criminal*, it was eyewitness testimony that most helped to create the "false alarm" (i.e., convicting the wrong person.)[17] In the early 1980s, a Virginia man named Marvin Anderson was convicted of rape based on eyewitness testimony, even though he had an alibi. The judge sentenced him to 210 years for the crime. He remained in prison even after another inmate confessed to the crime and provided details under oath in front of a judge. It was not until 2002 that Anderson was exonerated with DNA evidence that proved the other inmate committed the crime.[18] This case and hundreds of others like it show we value eyewitness testimony beyond its worth and are biased toward conviction. Before DNA analysis, those convicted on eyewitness testimony were all counted as hits, providing an inaccurate base rate, but now we have a more true picture of the signal detection grid, and it's an ugly truth.

Lieutenant Pigeon and Other Signal Detection Heroes

Signal detection is universal among living things. We humans just happen to have words and diagrams to describe it. This makes us think we are the smartest creatures on the planet, but that is only because we get to define "smart." Many animals out-do us in cognitive tests, including making mental maps (elephants) and combining more than one stream of information to make a good decision (rats).[19] When it comes to visual signal detection, humans are inept compared to the humble pigeon. It's not a fair comparison though, since pigeons can see far more gradations of color than humans can, including UV light.

Many people have heard of how homing pigeons were used to send messages, but in the early 1980s, the US military planned to use them for Coast Guard search and rescue in Project SEA HUNT. "How would it feel to be rescued after drifting at sea for days?," asked a Coast Guard magazine article. "Finally, a U.S. Coast Guard helicopter sees you and turns, flying straight to the rescue, and you, you are anxious to thank your rescuer – until you find out that *it's a pigeon!*"[20]

Based on mid-twentieth-century psychological research on animal learning, it was discovered that pigeons could be trained to identify shapes and colors. The pigeons of Project SEA HUNT had a chance to prove their worth in a test involving real helicopters and real floating buoys scattered across a large stretch of ocean.

Humans weren't great at finding objects floating in the water from a high altitude: their hit rate was only 38 percent. Pigeons, however, were 90 percent accurate. The differences between human and bird were even stronger on the first appearance of a buoy: pigeons caught it 84 percent of the time while humans only saw it 15 percent of the time. Pigeons demonstrated greater range than the humans, too, finding smaller buoys and targets further away, and in every case they were faster to see the targets than the human crew. The pigeons also had the advantage that they were not fatigued by having to remain vigilant. Even with hours between sightings, pigeons noticed the targets while human performance worsened over time. One Coast Guard officer still managed to get a dig in at the superior pigeons, saying, "They are too dumb to get bored."

Unfortunately, the technical reports on Project SEA HUNT did not include a usable false alarm rate for pigeons or human crew members. Many reports just said the pigeon's false alarm rate was "low" and the most complete publication reported the false alarms per search hour for pigeons (3.9), but no rate was reported for humans. (The calculation was complex. If a false alarm from a pigeon were also reported by a human, that false alarm was subtracted from the false alarms by the pigeons. This appears to mean that pigeons had 3.9 false alarms/hour *more* than the humans rather than 3.9/hour total. The thinking seemed to be "if the pigeons and humans both mistook another object for a buoy, it shouldn't count against the pigeon.") The probability of a hit was not reported per search hour so the two could not be compared.

The lack of data on false alarms was puzzling, especially since search and rescue is a case where false alarms have a high cost: they potentially direct the helicopter away from an actual target, use up search time, and cost flight resources such as fuel and person-hours. It appears to be another example of our natural tendency to focus on hits and misses rather than false alarms.

The Death of the Ball Turret Pigeon

In the end it was the impracticality of mounting pigeons in search and rescue helicopters combined with budget cuts and an ill-timed helicopter crash that sank the project. Since the birds were suspended in a transparent container under the helicopter, rescue was impossible after a crash landing, and the well-trained birds met their demise. Funding was pulled from the project in 1983. Now we have to depend on the tired, second-rate eyes of the human crew to find us when we are stranded in the ocean. No word yet on whether the TSA is funding the secret project LUGGAGE HUNT for improving airport screening.

Sensitive and Specific

In August of 2020 Magawa received the PDSA Gold Medal for bravery and devotion to duty.[21] A native of Tanzania, he was described as brave, friendly, and a determined worker. Over the course of four years, he found thirty-nine buried landmines in Cambodia. In a country with millions of mines there is still a ways to go, but experts like Magawa are on the front lines of mine detection. Also, Magawa is a rat.

A great attribute of mine-seeking rats is their weight. Although this particular species, African giant pouched rats, are *very* large for rats, they are still too light to trigger a landmine. They are also smart, have noses at least as discerning as the eyesight of pigeons, and are easy to train. Their training is a real rat race – they work about five hours each weekday on learning to discriminate the smell of a landmine from other scents, all for some mashed banana. They indicate finding a mine by stopping and digging with their little paws.

Magawa was trained by the APOPO, a Belgian non-profit with a mouthful of a name (the Anti-Persoonsmijnen Ontmijnende Product Ontwikkeling, which translates to "Anti-Personnel Landmines Removal Product Development"). The rats' training is extensive and their success is measured in hits, misses, false alarms, and correct rejections, with more available data than the Project SEA HUNT pigeons.

Why not just use a metal detector? The answer fits the theme of this chapter: the high cost of false alarms. The ground is full of non-explosive metal bits and trash and one estimate is that there are 1,000 false alarms per mine found using just detectors. The rats, meanwhile, can only graduate by demonstrating a 100 percent hit rate on four buried mines and no more than one false alarm. The rats have to be both *sensitive* and *specific*. The sensitivity rate is calculated as the number of mines they identified in an area created by their handlers divided by the number of total mines in that area. Their specificity is the number of false alarms per 100 square meters – a slightly different calculation than for other signal detection where specificity is the probability of a false alarm. It may be too much to ask of any single rat to be such a perfect performer, so typically every area is searched by a few rats and their combined scores make the final sensitivity and specificity rates for the team.

In a study led by psychologist Alan Poling, a professor at Western Michigan University, the trained rats were taken to Mozambique where they explored 93,400 square kilometers of mine-infested area.[22] The rats found 41 mines. Humans used other tools to check the area as well, like metal detectors, but found no additional mines. The rats succeeded with extreme sensitivity (100 percent hits and zero misses). But what about their specificity? In this real-world test, they had about 0.33 false alarms for every 100 square meters searched (617 false alarms), or about 39,383 fewer than would be expected from a metal-detecting human. My math is undoubtedly a brash generalization, as the mines aren't distributed equally and differ by country, but in terms of both hits and false alarms the rats are clear winners. For comparison, mammogram specificity is about 85 percent and their specificity is about 90 percent.

Ways to Test Signal Detection

It's just as hard to get the full picture of hits and false alarms in the dating world. How can Abby and Seth know whether the matching algorithms on dating sites are any good? Christian Rudder, one of the founders of *OkCupid.com*, had the same worry.[23] They were trying hard with their algorithms to match people well, because good dates are good for business. Their data bore that out – people were usually happy with their matches. Higher matches had longer conversations and were more likely to exchange contact information (ostensibly leading to a real, in-person date). But as Rudder said, "Maybe it works just because we tell people it does. Maybe people just like each other because they think they're supposed to?" It could be that they were creating matches via placebo just by announcing two people were a match.

This evidently worried Rudder enough that he decided to do an experiment. With 25 million users of his site, any well-designed experiment was guaranteed to turn up some reliable results. *OkCupid* created conditions where people showing a high match (according to the *OkCupid* algorithm) were randomly told either that they *were* a very

	Actual good match	Actual bad match
Told "good match"	20%	17%
Told "bad match"	16%	10%

Figure 8.2 Percentages of clients who exchanged a significant number of messages after being matched up by OkCupid.com.

high match (true) or that they were a very low match (false). The same was done for low matches: they were either told the site believed them to be a poor match (true) or a great match (false). The data they collected were how likely the matched pair exchanged at least four messages (Figure 8.2).

The first thing to note are the hits: even for good matches that were told they were good matches, only one in five exchanged four or more messages. Love is still hard to find! Also, as expected, odds for the actual bad matches told they were a bad match were even worse: one match in ten exchanged four messages. It only gets interesting when we look at the false alarms and the misses. With a false alarm, meaning the site told people they matched when they didn't, the conversation odds were close to when they actually *did* match. Clearly a placebo was at play, or the false alarm group would be much closer to the correct rejection odds. But love does find a way, even for the misled: those who were misses and were told they were not a match (when they actually were) also had fairly high odds of continuing conversation. Thus, they must have recognized something promising about the potential partner that superseded *OkCupid*'s judgement. A conclusion is that the *OkCupid* matching algorithms had value, but weren't perfect. The other conclusion is that we *do* like other people more when we think that we should because an algorithm tells us that we do.

Conclusion

We aren't good at considering false alarms when we think about success: we usually have a laser-like focus on the hits and avoiding misses. When a doctor detects something unusual on a mammogram, it's highly likely it is a false alarm.[24] But it is a rare patient who elects to forego a biopsy because the side-effects of the biopsy outweigh the possibility that the lump is malignant. We would rather go through the pain of the false alarm than miss a cancer diagnosis, even if it's extremely unlikely we have cancer. But hits and false alarms are tied together and there are penalties for increasing false alarms. Signal detection theory forces us to consider the hits, misses, false alarms, and correct rejections together instead of blindly adhering to the hit rate.

Signal detection is a hard-to-ignore framework for classifying the world once you know about it. We've moved from love, to nuclear annihilation, to the impossible task of the TSA, to cancer diagnosis, to pigeons fastened under Coast Guard helicopters, to crime and punishment, and back to love. There is a well-known cognitive bias that we "remember the hits and forget the misses." To this we can add, "and we don't often think about the false alarms and correct rejections, but we should!" I find great comfort in knowing that uncertain decisions can be quantified and strategy can be measured and adjusted to make decisions as good as possible.

9 APT PUPILS AND ALIEN INVADERS

Is this a dagger which I see before me,
the handle toward my hand? Come, let me clutch thee.
I have thee not, and yet I see thee still.
Art thou not, fatal vision, sensible
to feeling as to sight? Or art thou but
a dagger of the mind, a false creation,
proceeding from the heat-oppressed brain?
I see thee yet, in form as palpable as this which now I draw.
Thou marshall'st me the way that I was going;
and such an instrument I was to use.
Mine eyes are made the fools o' the other senses,
or else worth all the rest; I see thee still.

— William Shakespeare, *Macbeth*, Act 2, Scene 1

Daggers of the Mind

My introduction to philosophy came from my friend Alysonne during recess in third grade. She held up a tiny flower she found growing beside the sundial in the play yard. "What color is this?," she asked me. It was obviously purple and I said so. "I think it's purple too, but how do I know your purple is the same as my purple? Maybe what you call purple is green to me." An intense elementary school discussion followed about whether we could *know* we were seeing the same colors.

I'm sure it's a thought most everyone has had by the time they are an adult, but it was new and mind-blowing to an 8-year-old. How could we know? Could we ever know? What could you do to find out? Alysonne had a knack for introducing tough questions. It might be why, decades later, she's a lawyer. Only now I assume she knows the answer before asking the question.

"Do you see what I see?" brings with it a host of considerations for the engineered world. Mother Nature creates the colors of the flowers (or evolution does, to be specific), but humans choose the look of every watch face, dashboard, video game, headlight, and phone screen that exist in the world. On top of color, humans choose the brightness, the contrast, the saturation level, and hue. No longer a third-grader, I now know how differently each of us sees, due to myopia, presbyopia, colorblindness, just to name a few. Some people even have perfectly formed and working eyes, but are entirely blind because their eyes cannot communicate with their brain. Knowing how eyes send information to the brain, and how the workings can go awry, is crucial for making sure designs work for everyone and in a variety of situations, from first light at sunrise to fluorescent lights indoors to the dark of midnight.

Eyes send information to the brain, but it's up to the brain to interpret that I'm seeing a cheesecake on the table, and not an iceberg or just a picture of a cheesecake. Recognizing and differentiating objects is a big and important job for our brains, and we aren't perfect all the time. If someone mistakes a goat for a dog, no one gets hurt. Or confuses the word raven for craven. But if a military officer mistakes a caravan for a tank and calls in artillery fire, innocent people can die. Or if someone mistakes a car with one brake light out for a motorcycle, they may not give it enough room when passing. Thus, the question of how humans recognize objects (and what leads us to "recognize" them incorrectly) is a practical one, given how much control we have over the appearance of our world.

As a child, you see your first "A" and then you see a cursive A and then you see an *A* and so on, building templates for recognizing an A each time.[1] For complex objects, as a child you see a dog, and then you see a cat and call it a dog. But an adult corrects you, and now you have an instance for cat as well as dog. Seeing instances and recording them for later reference was the first theory that explained how we

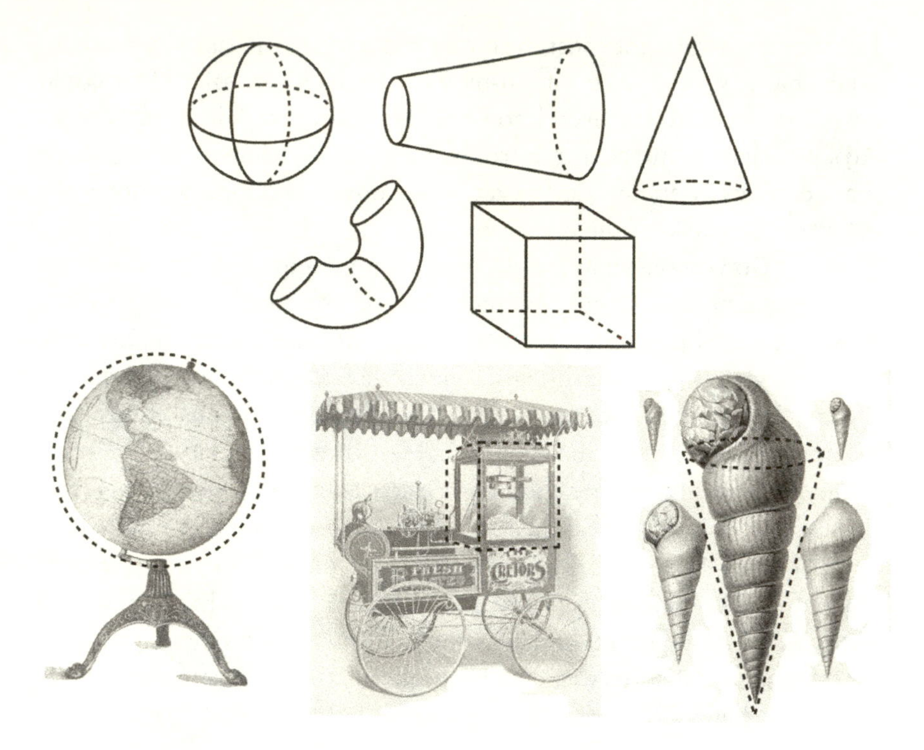

Figure 9.1 Simple 3-D forms called geons make up complex objects.

recognize. You're probably already thinking of the problem with this theory – the scale. Do we really have memories of every instance of every dog we've ever seen, from the strange little long-haired chihuahua to the blue-eyed husky? Have you really memorized what every person you know looks like from every direction? *If you've never seen the Magneto font, how could you read this sentence so easily?*

Unable to surmount the problem of scale, psychologists turned to another theory: that we recognize the features that build objects, called geons, and that combinations of those features make specific objects (Figure 9.1).[2] When you divide an object into features, a dog is a horizontal cylinder standing on four thinner cylinders, with a cylinder for its neck and one for its face. The tail can vary from the proud plume of the spaniel to the nub of the boxer, but *most* features of dogs have a "doggy" quality. If this theory is true, we should be more likely to confuse animals or objects that match on a high percentage of features, and indeed we do. The silhouette of a giant Bernese mountain dog might get confused for a bear, but never for an ostrich.

The Whole Is More than the Sum of the Parts

The first step in object recognition is to differentiate the object from the background. Imagine a cow in front of a cabin, wearing a bell. In your imagination you created the same visual cues that make the cow, bell, and cabin all separate objects.

Let's "look" at the cow. Can you see the cabin through the cow? I hope not. That's one cue the cow is in front of the house – it occludes the house. What about lines – the lines that make up the logs or windows on the cabin. Are they interrupted by the cow but continue on the other side of the cow? Same idea. The idea that interrupted lines should reconnect behind an object is our expectation that those lines continue (called *good continuation*) and show closure. What about the cow hide? Mine was brown and white. Yours may be another color, but it's likely different than the house or the grass. That similarity to itself (like the brown-and-white pattern present on the cow but nowhere else) is a cue that the cow was a single object. Yet another cue was proximity. The cow isn't scattered across the yard in pieces; the parts are near each other, meaning it's likely one object. (If your imagined cow was scattered across the yard, please see someone.) I didn't ask you to make the cow move, but if you did, all the parts of the cow would move together, which your eye detects as *common fate*. The word for all of these expectations that make us think an object is one thing and not part of another are called gestalts, a German word meaning "the whole is more than the sum of its parts (features)."[3]

These gestalts are applied in design, usually in ways that aid in correctly recognizing an object or objects. Similarity is often the key gestalt used in web interfaces, such as news sites (Figure 9.2). Proximity and similarity (the same grey color and font) group the menu items along the top into one object. No reader would confuse a menu item for a news story. Continuity means that the line across the top seems to continue through "The Human Factors Herald" as an underline and emphasis, separating the title from the news content. Even though the picture of the airplane is above the title below it, it's clear the photo belongs with the text at its *left*, thanks again to proximity. Finally, the list of articles on the right holds together as an object because of the similarity in font size and width and the close proximity of each headline to the thumbnail image beside it. Because of good gestalts, recognition of the different parts of the page from menu to headlines to sections

Log In

The Human Factors Herald

August 20th, 2021

Local | National | World | Sports | Editorials | Education | Technology | Health | Weather

BREAKING

Font size and boldness of headline shows importance

The proximity of this text to the headline above means they are connected. The changes in font and boldness above and below make it clear that this headline and text are connected to the photo to the right.

Because this photo doesn't line up with any of the other horizontal lines, it signals that it belongs with the text at left rather than any other text. Also this caption has proximity to the photo, indicating it must describe the photo.

The size and font of this means it is important

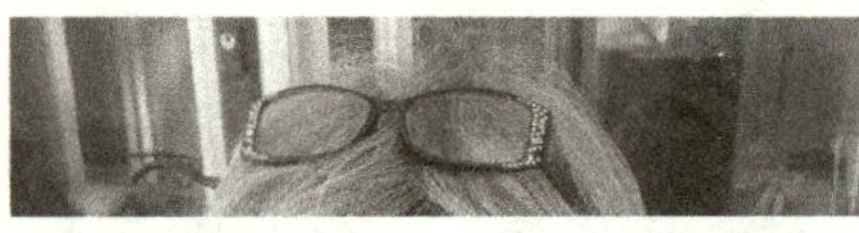

The size and font of this headline signifies it belongs with the others

The lines and whitespace make it easy to tell this text goes with the headline. No confusion thanks to Gestalts.

It is easy to tell that this is the actual text of the article, while the text above is a brief summary because the font size and color changes between them. The text would continue below the "fold" - and it's especially helpful if part of it is cut off, so the reader can tell it must continue if they scroll down the page. I cut it off in this example to show

Local

Grey background indicates this story is special

Perhaps it is the most recent or most important local story.

Good continuation

Even though the line to the left doesn't connect with those above and below, it's clear they are a common divider.

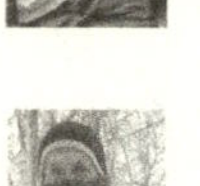

Similarity

Because all stories in this column share font and color, it's clear they all fall in the Local category.

Another local story

Cut off text visually helps the user know to scroll to see the rest of

Figure 9.2 Typical layout of the front page of a news website showing how gestalts allow for easy grouping of sections, menus, and headlines.

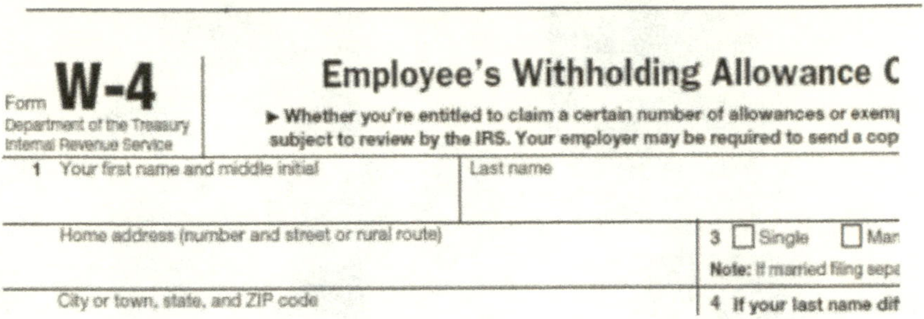

Form **W-4**
Department of the Treasury
Internal Revenue Service

Employee's Withholding Allowance C
▶ Whether you're entitled to claim a certain number of allowances or exem
subject to review by the IRS. Your employer may be required to send a cop

1 Your first name and middle initial | Last name
Home address (number and street or rural route) | 3 ☐ Single ☐ Mar
Note: If married filing sepa
City or town, state, and ZIP code | 4 If your last name dif

Figure 9.3 US Tax form with conflicting gestalts. On which line does the address go?

is unconscious and effortless. The effort saved can be dedicated to reading and understanding the news of the day.

But sometimes gestalts are misapplied and we are led to mistakenly group items into the wrong objects. I see this most commonly in forms, where the label for a field is just a tad too far away from the textbox, leading me to write my address in the space for my name, or my postcode in the spot for my street number (Figure 9.3). The line meant to signal a different box on the form ends up looking like a signature line, inviting me to write on it.

It doesn't take much change in a gestalt to signal what is an "object" and what is not. Figure 9.4 shows two attractive layouts for pairing an image with a caption. The photo with crossed arms is Annika, but in the left panel (1) the proximity and intermediate object (the horizontal line) makes it appear Annika is the rock climber. The right panel is a small revision to the layout, but it becomes evident which photo is of the explorer and which of the adventurer. In short, gestalts are strong and unconscious signals that are great when used well and difficult to overcome when used poorly.

Getting the Jizz

Though there seems to be some general process we use to recognize most objects, there are some things that we've seen so much and learned so well that we are perceptual experts. For most people, faces fall into this category. We're so good at recognizing faces that we see them even when they aren't really there, such as in the bark of a tree or in the burn marks of a McMuffin sandwich. But for the highly

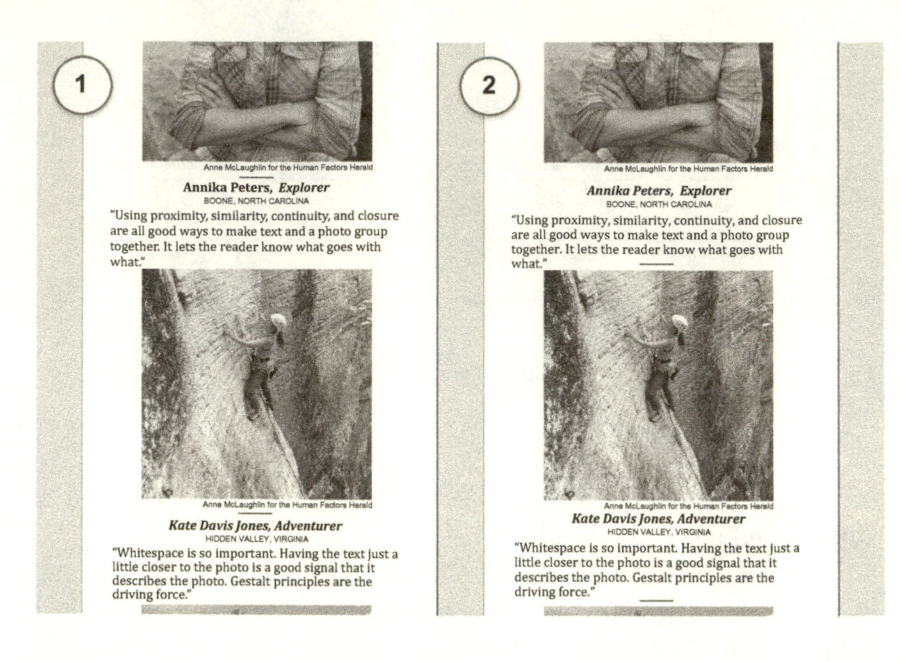

Figure 9.4 Does the text for Annika go with the photo above or below? It probably depends on whether you consult layout (1) or (2). In (1), gestalt principles suggest below, because it has closer proximity to that photo and less white space between the photo and text. But the text actually describes the photo above it. In (2), the white space and lines correctly suggest the photo and text go together as one object.

trained there are other areas of expertise: car buffs, birdwatchers, and radiologists come to mind. Once someone can recognize subtle attributes in the blink of an eye, with little feeling of effort, they've developed "perceptual expertise." They've changed their object recognition for a few specific objects from the slow, effortful, and error-ridden process of a novice to the fast, accurate, and automatic process of an expert. No matter what the subject matter, these experts report similar feelings when they recognize one of the objects in their expertise: almost a shuddery "bink" in the brain. Birdwatchers have a name for this. They call it "getting the jizz."[4]

For psychologists who study perceptual expertise, birdwatchers and radiologists are their favorite participants. Radiologists are favored because they have standard training and there is obvious practical benefit to knowing how they acquired and refined their skills. Birdwatchers are favorites because there are a lot of them and they seem inordinately interested in participating in research studies for free. One of the most interesting findings from studies of birdwatchers and radiologists is that experts are faster at recognizing specifics than generalities, the opposite of the nonexpert.[5] Personally, I don't know much about birds. I recently spent a week puzzling out why the baby birds outside my window didn't look like their mom, a junco. (They were starlings, with mama junco the victim of a nest invasion.) If you showed me pictures of animals, I would be much faster to say "It's a bird" than I would "It's a junco" or "starling." For birdwatchers, this is reversed. They are faster to say "Chestnut Cheeked Starling" than they are to just say "bird." They almost can't suppress the "Chestnut Cheeked Starling." It just leaps from their mouths. You can see this expert reversal effect in many domains, from car-nuts to architects to botanists, all getting the jizz. But this expertise isn't limited to hobbies – if you're reading this book you have high expertise in recognizing words.

The Fools o' the Other Senses

When we skim letters, words old and new, our brain processes them much faster than we can say them aloud. The lay explanation is that we see words with our eyes, they send the message to our brains, and our thoughts are the outcome. But the details of that process are devilishly counterintuitive and interesting.

Let's go back to the organ that touches the outside world and starts that internal journey to become information: the eye.

Eyes seek out information. We explore our environment by moving our eyes, our head, and our body to feed our senses. I spoke with Rick Tyrrell, an expert in perception at Clemson University, who described the loop of perception and feedback that endlessly occupies our waking life: "We need to move in order to perceive. The tiniest eye movement is exploration, as are countless head and neck shifts, or peering around the tall person blocking our view, or touching the fabric of a piece of clothing in a store. Movement is so important but so underappreciated."[6] "Optic flow" from movement and perception informs us about our world, while coordinating our muscles to act. Our actions inform our perceptions and our perceptions guide our actions.

What we see, compared to what is actually present in the world, is due to the types of receptors in the back of the eye, on the retina. These receptors respond to light and fire signals back to the brain, transforming the energy of the photons in the outside world to the internal language of our neurons.

We have two kinds of receptors: rods and cones. One looks a little like a cylinder (rod) and other a little like a funnel (cone), so the names actually make sense.

Cones are tightly packed in the middle of the retina, a thin film on the back of your eye. The cones need good light to function. In return for demanding good light, they show us the colors of the rainbow and the fine detail of lines and letters. But don't be too complimentary of the cones, because they miss a lot. For one, with only three types of cone color receptors, there is a limit on how many colors they can differentiate. Many animals have better color vision than humans thanks to the kinds of cones they possess. Bees and some birds can see ultraviolet light and the extra colors connected to it. Pigeons are especially great with colors, and can differentiate between shades of yellow we don't even know exist. Even tiny mantis shrimp can see colors that we cannot. But with our measly three cones we can see millions of colors. Well, unless you are one of the 8 percent of men and a half-percent of women who lack one of the usual types of cone, leaving you with "red/green" colorblindness, the more rare "blue/yellow" colorblindness, or the most rare, full colorblindness, where the world is only black, white, and gray. (Given these statistics, it seems diabolical that so much of the world has decided on red/green for stoplights.) If you are not colorblind and lucky enough to reach age 65 or so, don't be too smug. We lose some of our ability to detect colors as we age because the lenses in our eyes yellow, making blues harder to differentiate from each other and from green.

The responses of the rods and cones eventually have to wind their way back to the brain. How they do this is very strange and the opposite of any sensible wiring schematic.

The retina sounds simple. Lay a film of receptors across the back of the eye, detect the light, and boom: a camera. Also like a camera, the image that falls on the receptors is upside down (due to the double convex lens) and has to be turned "right side up" by our brains. Ok, that's an extra step it might be nice to skip, but we can manage. But there is an even stranger and more backwards structure in the eye. In *front* of the receptors are multiple layers of other cells, cells that the light has to go through to reach the retina. Those cells don't react to light at all. They are just messengers, turning the light information from the rods and cones into data to be processed by the brain. If they were behind, all the light could fall where it should on the retina, and then work back toward the brain. By being in front, these cells are entirely in the way. Further, we have an optic nerve that takes up some of the real estate in the retina where we don't have any photoreceptors. This gives us a blind spot. We should see a cobweb of shadows on the world, since these cells are always blocking some of the light.

We don't see the cobweb or the blind spot because the brain is aware of these shadows and blankness. Since they have been there a while (since birth), the brain learns to ignore them and fills in the world that should be there. Your eyes are constantly making tiny movements, even when you believe you are holding them still. These movements help fill in the spots behind the cobweb of cells and the blind spot the same way you can see through a spinning electric fan.

So, to summarize: Your eyes get an upside-down image that is automatically flipped by the circuitry traveling to your brain. While doing this, they have to fight a covering of many layers of cells blocking the light-sensing parts of the eye. Eyes do this by moving constantly and combining light received at different times into one "picture." After these acrobatics, the brain finally gets to try and interpret this mess into the world around us.

There are animals that evolved more sensible eyes. Squid and octopi have "camera-like" eyes just like we do, but with the layers of cells tucked behind their retinas instead of in front. They don't even have a blind spot. In a more efficient world, we'd have eyes like theirs. But our eyes work well, even as backwards and clunky as they seem to be. We can take note of our limitations and design around them to make the world a safer and less frustrating place. By knowing how the eye works we can create designs in the world that work with our visual capabilities and limitations, rather than against them.

Visible Muscles

The pupil, that little black hole in the center of the eye, is controlled by the attractive iris. Incidentally the iris is the only visible muscle with no skin covering it. The pupil changes size for a multitude of reasons. The most obvious is light. The pupil constantly monitors the environment, adjusting quickly to let in more or less light to keep the retina from being oversaturated or starved for input. There are limits, of course. You can still damage your eyes from too much light in a sunlit snowscape. Our eyes adapt in dim light by enlarging our pupils to grab as much of the waning light as possible, eventually maxing out. If a light goes on, they instantly respond the other way, tightening the pupil to restrict the harsh incoming light on the now-sensitized receptors, until they can adapt.

The second reason is also fairly well known in modern culture: drugs. Some drugs turn the pupil to pinpoints – letting in very little light. I remember the first person I met with eyes like this. He spent forty-five minutes emphatically lecturing me on rock-climbing technique and backing me into a picnic table. It was only later that I realized, from his tiny pupils, that maybe he wasn't just really into rock climbing. Alcohol does the opposite to the pupil. It relaxes the muscles in the iris and makes them respond more slowly. This is one reason why police officers shine a light into the eyes of suspected drunk drivers. They can tell when left and right pupils don't match in size or when a pupil is responding slower than it should to a sudden light.

The next pupil changer is of great interest to anyone wanting to measure effort. The harder we are working, attentionally speaking, the larger our pupils get, even if light does not change. A math problem like $5 + 4$ elicits very little pupil change, but $3(4 \times 4)/(2-3)$ will make pupils get a little bigger. Solving simultaneous equations will make them get much bigger. How can we use this knowledge? Imagine you have two potential displays for a fighter pilot to use. They don't make any errors with either, and the two options seem to be similar in most ways. But since you can't test pilots using the system in every situation, you can measure their pupil changes with an eye-tracking machine and see how much effort each option requires. The results might suggest one interface design takes less effort from the pilots than the other.

The last reason for pupil change allows mindreading. Your pupils get bigger when you see something you like. It is involuntary and uncontrollable, leading poker players to wear shades that hide this "tell." It is also why people ill-advisedly used to put belladonna, from the nightshade plant, into their eyes a hundred-plus years ago. Belladonna ("beautiful woman") is a poison that dilates the pupils, making the woman who poisoned herself even more beautiful. Why are dilated eyes more beautiful? Because we like people who like us, and dilated eyes make it look like the other person is *very* interested in us. You can still tell what people like by putting different items (or people) in front of them and using an eye tracker to measure changes in pupil size. For those with romantic partners, you can try exposing a part of the body your partner should find of interest while watching their pupils. My bet is on dilation.

To sum up, pupil size adjusts due to effort, interest, and light. When the pupil is large, but still not getting enough light to see, eyes begin to adapt to the dark. The first change is that it gets harder to see colors. Reds start to look blue, dark colors start to look light, and what was bright starts to look dark.

If you want to feel how powerful dark adaptation is, try this self-experiment. At night, while reading this chapter (or doing anything else in the light), cover one eye. You could use an eye patch or just solidly hold your hand over an eye. Hold it there for about thirty minutes. Then, with it still covered, turn off the lights so that you're in a dark space.

Open both eyes.

I guarantee you will worry that you've gone blind in one eye. It is an odd feeling, and one I highly recommend! The eye that was reading in the light seems suddenly struck blind, because the other dark-adapted eye can see quite well. The feeling is so strange that it was reported in the news that several people went to the eye doctor for "temporary blindness."[7] They had been using smartphones in bed with one eye pressed into the pillow, so one of their eyes adapted to the dark. When they turned off the lights, the eye that had looked at the smartphone seemed violently and suddenly blind compared to their eye that had grown used to the dark. You can also use this technique for going to the bathroom in the middle of the night. Cover one eye when you turn on the lights, then cover the other eye when you turn them off again so you can see your way back to bed in the dark. Don't say you didn't learn anything useful from this book.

Full dark adaptation takes about forty-five minutes, but depends on how bright the starting point was and how dark the endpoint is. Moving from the sunny outdoors to a heavy curtained room makes adaptation take longer than moving from the sunny outdoors to a shaded hallway to a darker entry and finally into a curtained room. Some airports and other spaces are designed to aid in dark adaptation, with moderately lit liminal spaces between the darkest and lightest sectors. Hartsfield-Jackson Airport in Atlanta has a particularly beautiful example of this, with a greenery-covered tunnel as a liminal space between gates, complete with birdsong and the occasional computer-generated cloud peeking through the leaves.

The Hunt for Red Toyota

Dark and light adaptation should also influence the interfaces we need to use in dark and in light, such as dashboards. This is why red lighting is the norm on submarines. Red light is on a wavelength that least harms dark adaptation, so red light is used when we want to see as well as we can, but not lose our dark adaptation. Next time you're in a car, check the lights on the dash in front of the driver. What color are they? They should be red to preserve the driver's night adaptation, but they probably aren't. In my unscientific survey of "cars I've been in across twenty years," only my 2003 Toyota Matrix had an all-red dash. Even Toyota didn't maintain this, with newer models having blue or even white in the dash. At that point, the best thing you can do is to dim the dashboard to keep your night vision. Even in my Matrix there was an aftermarket cruise control stalk with a green LED. Every time I used cruise control, I thought about what a terrible design decision this was. Not only was it bright and green, making it the most eye-catching object in a dark car, it alternated between visible and hidden, depending on the turn of the steering wheel. This made the bright little light appear to blink, the most attention-grabbing action a display can do. Using cruise control at night was both tiring and distracting. Thanks, Toyota.

Using Evolution for a Revolution in Safety

In 2018 over 6,300 Americans died from being hit by cars while walking, running, or standing. The number is tracked by the National Transportation and Highway Safety Administration and it's risen each year for at least the last ten years. More than two-thirds of these accidents happened at night, even though far fewer people walk around in the dark compared to daytime.

One of our most human attributes, our vision, makes nighttime a danger to us. But another core attribute, our social nature, offers hope for designing technology to overcome that danger.

We are diurnal creatures, adapted to live and hunt during the day. This makes us fairly unique as most mammals are nocturnal. Indeed, we are an extreme of the daytime mammals, as primates and monkeys are the only mammals to have evolved eyes that work as well in daylight as birds and reptiles. We can adapt to the dark, but cannot compare to the real creatures of the night, like owls and raccoons or even dogs. Our second core attribute is that we are social creatures, finely tuned to the emotions, actions, and likely movements of other humans. We can identify mood at a distance, sense friend from foe, and predict where and how fast a person might move. It turns out we can use our core social nature to overcome our poor night vision.

You might be thinking light is the answer. More street lights! Better headlights on cars! Turn the night into day! In some places, such as a pedestrian square or a busy intersection, that's a viable solution. Modern automobile features, including headlights that match the steering angle to illuminate turns and automatic switching between high and low beams, help us see at night. But since we can't depend on light to follow us everywhere, we need to make it so the little light we have is maximally helpful.

Joanne Wood, an optometry professor at Queensland University of Technology, and Rick Tyrrell, the aforementioned human factors psychologist at Clemson University, have spent many years using the basics of vision to improve recognition of the human form at night. Their findings were surprising and illustrated the vast gulf between what we think is visible and what's actually visible.

Their tactic was to start with the essentials of an object, those shapes and movements that are instantly recognizable, and devise technology that mapped them onto the human body. They applied reflective material to a person's ankles, knees, wrists, and elbows. In the dark, those patches were all that were visible. When they showed just the reflections, particularly on a moving person, viewers could easily tell it was a human being from the *biomotion*. And it's not just us. Lab studies showed that cats recognized other cats and pigeons recognized other pigeons just from dots at the major joints. Even human infants respond to biomotion.

The question that these human factors researchers asked was whether applying biomotion markings could help drivers recognize nighttime bikers, walkers, runners, and road workers. Reflective and fluorescent vests were already common, but reflectors on the torso don't give much biomotion information. Too often, when drivers encounter a pedestrian wearing a vest with reflective markings on it the drivers detect a "thing" but fail to recognize what the thing is. Would biomotion markings give drivers a better heads-up?

Drivers recognized pedestrians at distances orders of magnitude farther when reflectors were mounted to the pedestrian's extremities,

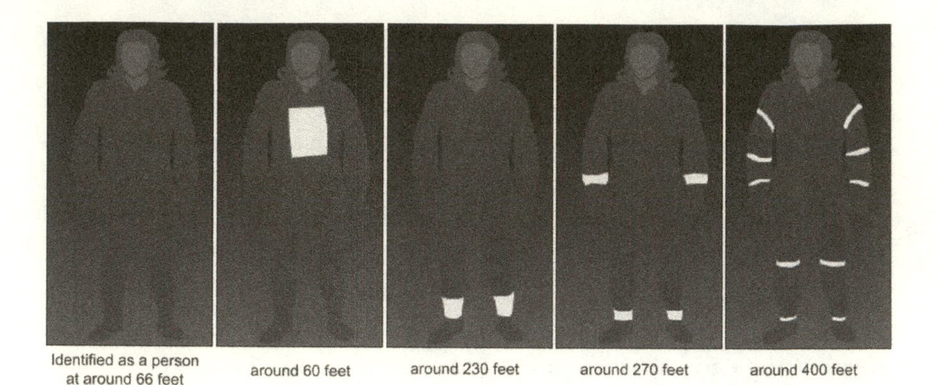

Figure 9.5 A dark-clothed pedestrian wearing the same amount of reflective material in different locations. Notice there is no real difference between nothing and a vest. All the improvements come from highlighting biomotion, the more the better.

even when the total amount of reflective material was held constant (Figure 9.5).[8] *There was no real difference between wearing dark clothes and wearing a reflective vest.*

To give you an idea of what a huge impact the reflector location had, a walker wearing just ankle and wrist reflectors, no vest, was identified from *four times* farther away than a reflective vest. A walker wearing full biomotion stripes was seen *six times* farther away. With studies like these it is hard to understand why the only advice on the Centers for Disease Control (CDC) website is to "Increase your visibility at night by carrying a flashlight when walking and wearing reflective clothing, such as reflective vests."[9] Ankles and wrists should at least get a mention.

Tyrrell told me about the first experiment he and Joanne Wood ran on real streets with real drivers and pedestrians:

> We put pedestrians out there on a closed road and changed their clothes to see how good drivers were at noticing them. We tried low headlight beams, high beams, and different kinds of clothing combinations. When the pedestrians wore no reflective material the drivers were horrible at seeing and recognizing people, even worse than we had suspected.
>
> In our first big study we were in Australia on a closed road in a neighborhood at night. It was a real road, with curves, hills, normal intersections, and proper signage. We carefully

> positioned experimenters on the roadside and asked them to walk in place while volunteers drove past them. Drivers just had to slap this big metal plate in their car as soon as they saw a pedestrian. I mean, in this study the drivers had every advantage. They knew they were in a research study. They knew there would be pedestrians around and they knew we were measuring their ability to respond to the pedestrians. There were no surprises. This was really a best-case scenario for the drivers to see the pedestrians.
>
> When we put the pedestrians in dark clothes, they were basically unrecognizable. In our worst-case condition, drivers were using their low beams, were facing another vehicle's headlights, and approached a pedestrian wearing dark clothing. In that situation, 95% of the drivers never responded to the pedestrian. Even when we took away the glare from the opposing headlights the drivers didn't see the pedestrians until they were right up on them, about 18 feet away when going 35 miles an hour. Way too late to be safe. But when we put reflective bands on their ankles, knees, wrists, and elbows, the results flipped. Suddenly, the drivers were identifying pedestrians from more than 450 feet away. Plenty of time to brake.[10]

Another study they did was similar, but with an intriguing twist. They asked volunteers to walk on the roadside and mark when they thought that the approaching driver could recognize *them*.[11] Pedestrians clicked a handheld button as soon as they thought the driver could see them. After all, if we overestimate how visible we are maybe we're more likely to step out into a road or run alongside the curb at night.

The researchers had good reason to think this based on everything they knew about the eye. Pedestrians' eyes are probably more adapted to the dark than the drivers' eyes, so the world doesn't look as dark to them as it does to the driver. Also, headlights are incredibly intense and impossible to miss. Maybe a pedestrian seeing bright headlights approaching thinks that it's just as easy for the driver to see *them*.

Pedestrians greatly overestimated how visible they were to the drivers. When wearing dark clothes, pedestrians clicked the button when the car was almost 300 feet away. But the driver didn't recognize the pedestrian until they were only about *18* feet away. The pedestrians also overestimated their visibility when they wore white clothes and when they wore a reflective vest. It was only when the pedestrians wore the biomotion stripes that they underestimated their visibility. (So it's not surprising that pedestrians don't typically wear biomotion stripes – they don't realize how useful they are!)

I asked Rick Tyrrell how he felt about knowing how effective biomotion stripes are, but only rarely seeing people use them. "I've only ever seen vests at night-time construction sites," I said, "and I've never seen a jogger or biker with wrist and ankle reflectors." He laughed, as he had clearly been asked this question before.

> I'm doing my best to tell people about the benefits of biomotion, but it's an uphill climb. One of my fantasies is to get all this information about biomotion and how helpful it is into high school biology textbooks. Our brains have built-in circuitry that is exquisitely fine-tuned to recognize the motion of other humans. It's biology. Think about how useful it would be if more people learned about it.

It is our social nature and dependence on interacting in groups that has led to us evolving this perceptual sensitivity. I agreed this explained why babies and birds could do it, but I'm reserving judgement on the notoriously aloof cats.

Tyrrell and his collaborators did have one big win. As a result of this research, road workers in Australia and New Zealand are now required to wear biomotion reflective gear at night.[12]

Dark clothing is like an invisibility cloak at night. There's just not enough contrast. In some cases, a reflective vest doesn't help at all – drivers see an unidentifiable shape floating at an unknown distance. Seeing "something" isn't enough – the form has to be recognizable. By capitalizing on our perceptual sensitivity, biomotion markers make the nearly impossible job of recognizing a dark figure at night easy and effortless. Yet the USA (as of early 2021) does not regulate the use of biomotion for people working at night and doesn't even include it in safety recommendations from the CDC or transportation safety organizations. Just today I looked through a list of products for the "avid runner." A headlamp was included, but no mention of ankle or wrist reflectors. But now you know.

Of course, it's one thing to recognize a figure as human. But there is a whole separate process in the brain for recognizing a *particular* human.

YOU'RE NEXT! YOU'RE NEXT!

One of the ways we understand object recognition is by studying those in which it has gone awry. After a car accident at age 22, "David" stopped recognizing his own parents.[13] Neuroscientist V. S. Ramachandran presented David as a case study in a documentary on Capgras syndrome for NOVA. He described David as

> articulate. He was intelligent, he was not obviously psychotic or emotionally disturbed. He could read a newspaper. Everything seemed fine, except for one profound delusion. He would look at his mother and he would say, "This woman, doctor, she looks exactly like my mother, but in fact, she is not my mother. She is an imposter. She is some other woman pretending to be my mother."[14]

The same was true for David's father, all imposters identical to the originals. Susannah Cahalan, author of *Brain on fire: My month of madness*, experienced similar symptoms brought on by brain inflammation.[15] Other causes include stroke and certain types of dementia.

I was reminded of Jack Finney's brilliant 1954 book, *The body snatchers* (made into the 1965 film *Invasion of the body snatchers*). I can't say for sure whether Finney knew someone with Capgras syndrome, but I'd be surprised if he didn't. Here is an excerpt from the book:

> Wilma sat staring at me, eyes intense. "I've been waiting for today," she whispered. "Waiting till he'd get a haircut, and he finally did." Again she leaned toward me, eyes big, her voice a hissing whisper. "There's a little scar on the back of Ira's neck; he had a boil there once, and your father lanced it. You can't see the scar," she whispered, "when he needs a haircut. But when his neck is shaved, you can. Well, today – I've been waiting for this! – today he got a haircut – "
>
> I sat forward, suddenly excited. "And the scar's GONE? You mean – "
>
> "No!" she said, almost indignantly, eyes flashing, "It's THERE – the scar – exactly like Uncle Ira's!"[16]

Of course, in the book Uncle Ira really is an alien, but Wilma is a Capgras sufferer if I've ever heard one. In John Carpenter's *The thing*

(1982), Antarctic scientists and workers are picked off slowly by an alien that perfectly mimics the unfortunate victim, only to emerge with aplomb (and an impressive amount of blood) when threatened. The paranoia these situations generate for the other characters is what makes them so engaging. This plot device has become so common in books and film that it has a name: "kill and replace."

Other than generating great movie plots, Capgras keyed us into a critical component of recognition: emotion. For most objects, there isn't too much emotion tied up in recognition – a rose is a rose is a rose. But with some objects and the people who are closest to us, we need emotion to recognize them. You don't just see or hear your loved ones, you *feel* them. Our best explanation of Capgras is that the emotional line to the recognition center of our brain (in the amygdala) is cut, while the visual line remains. "She looked like my mom, but she wasn't my mom" is a common complaint of the Capgras sufferer, despite every hair and freckle being in place on mom. They aren't able to reconcile what they see with a lack of feeling, and thus come up with a "sensible" solution: "She's been replaced by an alien (or a demon or whatever is culturally extant)."

There are some fascinating limits on Capgras syndrome. In the case study of David, his lack of recognition was limited to a disconnect between vision and emotion – he readily recognized the voice of the loved one he just visually dismissed as an imposter. If the "imposter" left the room and called from the hallway, David fully believed his mother/father was outside the door. David's parents discovered this and used it, as said by David's mother: "We got so tired of him saying you're not my dad, you're my dad, you're not my mother, you're not my mother. We decided, ok, you go downstairs, call, on the phone, and say 'David, hi,' and on the phone he would know he was his dad. On the phone he never ever had this problem."[17] From this and from a comparison to Capgras sufferers who only *heard* imposters rather than seeing them, we learned that the auditory and visual pathways to the amygdala were independent.

When David's parents started calling him to convince him of who they were, they designed a new system of communication. One that worked with David's limitations rather than against him. This shows off two of our best human qualities, the ability to adapt and the ability to create tools to solve problems. David's issues with recognition are on a personal scale, within his family and shared with the few other Capgras sufferers in the world. But the solutions his parents developed are just as good an example of the power of good design as any red-lit dashboard in a Toyota.

Conclusion

Are the eyes the windows to the soul? Perhaps. Are they the windows to the brain? Yes, figuratively and literally. We can recognize things in our world through touch, hearing, and smell, but vision is king, often dominating the other senses when it is available. But no image falling on the eye is enough for recognizing objects around us – the brain has to translate the image into information we can use about identity, location, and movement. Knowing this can be used in the world we create through engineering and design. We must account for the fallibility of our vision. Designs can work with us or against us in this, either helpfully outlining the biomotion of road workers or tiring our night vision with beautiful bright-blue dash displays. The choice is ours in what we create and legislate, but also what we are willing to accept when we purchase products.

10 A RELATIVE TO TRUTH

Memory is a complicated thing, a relative to truth but not its twin.
– Barbara Kingsolver, *Animal dreams*

In 2006, Lenny LeClair went in for abdominal surgery. The surgery went off without a hitch. Or seemed to.

However, when he was interviewed by *USA Today* in 2013 he recalled the horror of his post-surgical life. "I started throwing up," he said. "Twenty-four-seven."[1] When he complained to his doctor, he was prescribed milk of magnesia and told he would be fine. But everything was not fine – surgical sponges left in his torso were slowly working their way into his intestines. The sponges were left despite the nurses counting the surgical equipment before and after the operation. LeClair was far from alone: a surgical implement is *retained* (left inside a patient) in about one out of every 7,000 surgeries.[2] Counting the surgical instruments in the room before and after surgery is encouraged by the World Health Organization, but not mandatory.[3]

"Over about 5 months I lost a hundred and sixteen pounds," said Lenny. As the sponge rotted and pierced his colon, he felt worse and worse. LeClair's family checked him into a hospital where they finally discovered the sponges. "[The doctor] said if I'd have waited three more days, I wouldn't be here. I've had six surgeries, and have two more to go." Due to the extensive damage, LeClair will have a colostomy bag for the rest of his life. Remembering a sponge sounds so simple. Counting items before and after surgery sounds so simple.

How could healthcare professionals forget items inside patients again and again?

***

In June of 1998 on the coast of Australia, a fisherman came across a square piece of chalkboard. Written on it in grease pen was a message, reading, *[Mo]nday Jan 26; 1998 08am. To anyone [who] can help us: We have been abandoned on A[gin]court Reef by MV Outer Edge 25 Jan 98 3pm. Please help us [come] to rescue us before we die. Help!!!*[4]

The board was a scuba diver's slate, owned by Tom and Eileen Lonergan. The couple was accidentally abandoned by their dive boat months earlier and were never found. They survived long enough to write the message found on the slate – at least 17 hours in the water. If this sounds like a one-time nightmare of a freak event, it's not. In 2011, Fernando Garcia Puerta and Paul Kline were left in the waters near Miami but fortunately spotted by a passing yacht.[5] The same year in Australia, Ian Cole was left behind, but managed to swim to another boat for rescue.[6] Daniel Carlock was abandoned off Newport Beach by his dive boat in 2010, but was lucky enough to be seen by a troop of seafaring boy scouts.[7] Another diver was not so lucky – she was never found after a dive boat left her off the coast of California in 2015.[8]

There is no official count of how many divers have been left behind by their dive boats. The stories here come from just the reports that made the news. But just like the surgical sponges, it's hard to imagine how a boat full of divers and the company that controls the boat forget the *people*.

***

I have my own story of memory failure, a quotidian counterpoint to the serious tales thus far. On August 25, 2015, I forgot to go for a run.

Up to that point I had a running streak of four-and-a-half years, running at least a mile every single day. I hated running, but the streak was a strong motivator for me (a sedentary, loss-averse psychologist). The longer the streak went, the more I cared about keeping it going. The streak was part of my identity, something I thought about multiple times a day and had kept alive through challenging circumstances: undergoing general anesthesia, sprained ankles, running in dress shoes after a too-long day in Washington, DC, and laps around the Madrid airport en route to a conference in Europe. This was one case where falling prey to a logical fallacy (the "sunk cost fallacy") did some good in keeping me healthy. I'd do anything to keep from breaking the streak.

I always thought an injury would take me down, either an ankle I couldn't hobble a mile on or maybe an emergency appendectomy. But on July 26, 2015 while on vacation in South Lake Tahoe, my eyes sprung open at three in the morning. My first thought was "I forgot to run yesterday." After 1,580 days of remembering, I forgot.

We talk about memory as if it were a thing; either you have a memory of an event or you don't. "A memory" seems like a physical entity that resides in your brain. Sometimes it's harder to access than others, like when you can't quite remember someone's name or where you met them. And sometimes, when a memory is an intention, it seems to disappear and you forget what you were looking for when you came into the room, or forgot to take a certain medicine on time, or that the two divers you talked with that morning aren't on the boat. But one of the most fascinating advances in psychology in the last few decades was the discovery of the physical representation of memory, the act of remembering, and forgetting. And it's nothing like we assumed.

Bridges across the River Lethe

> Lethe, the river of oblivion, rolls his watery labyrinth, which whoso drinks forgets both joy and grief.
>
> – John Milton, *Paradise Lost*

The single most surprising findings about our memories is that they are not static cells or clusters of cells. Unless they are being formed or recalled, they don't exist anymore than a song exists on an unplayed piano. Neurons are the foundation of memory, but no one has a neuron that represents her grandmother or the smell of a new car. Instead, when memories are formed or recalled, they create connections to each other, and these bridges are the memory. Each time a memory is recalled, the bridges between neurons are changed, rebuilt just a little, meaning that no matter how similar a memory feels from one day to the next, every time you remember it, it is slightly different. It is the pattern of activation, bridging between neurons, that makes a particular memory. It also explains why memory is so fragile and subject to change (yet the person with the memory feels it is unchanged). We know these bridges, made of proteins, create memory because we can prevent those bridges from forming with a drug (in mice). When the bridges don't form, the memory doesn't either.[9]

A Pastanalogy

Think of memories like strings of spaghetti. To form the spaghetti into a particular shape, you boil it, and while it's soft twist it into whatever memory you want. When it dries, it's solid. But every time you remember it, you have to boil it again, and when it is in that state, the shape can be revised, reformed, or even obliterated.

Memories are most susceptible to change when they are being formed and when they are being remembered. This is because the same process that formed the memory is active during remembering, making those solid bridges of dried spaghetti more flexible.

We're All Noodlebrains

The dangers of soft spaghetti can strike even the biggest brains. In 2008, astrophysicist and all-around genius Neil Degrasse Tyson gave a speech where he quoted President George W. Bush as saying, in regards to the terrorist attack of 9/11, that "Our God is the God who named the stars."[10] Such phrasing would seem aimed at pitting Christians against Muslims (sharing the God of the Old Testament notwithstanding). Indeed, Bush did say a similar phrase. But he said it two years after 9/11 when mourning the *Columbia* explosion: "The same Creator who names the stars also knows the names of the seven souls we mourn today." By 2008 the two speeches had melded (reconsolidated) in Tyson's memory.

Tyson is likely weary that his misstep is pulled up again and again, and I only hope he appreciates that it is his otherwise exemplary intelligence that makes him such a good example. We all experience reconsolidation memory errors, but rarely is there such public evidence to prove us wrong. He is also a good example of how we should handle being confronted with evidence that our memories were incorrect, eventually saying, "And I here publicly apologize to the President for casting his quote in the context of contrasting religions rather than as a poetic reference to the lost souls of Columbia. I have no excuse for this, other than both events – so close to one another – upset me greatly."[11]

These false memories happen to all of us. If you're thinking that you've never had a false memory it is only because your own false memories seem so true that you have trouble identifying that they are false. An example from my own life happened just a few weeks ago, and I only know that my memory was false because incontrovertible evidence was presented. It still *feels* true.

I was recently back in my hometown helping my mother to clean out and donate some of my late father's clothes. We came across a beat-up yellow flannel shirt, which I remembered vividly from my childhood. Looking at that shirt, I clearly saw my father dressed in it to hunt quail. Camouflage pants, suspenders, big heavy boots, and a pair of fairly unflattering dark-rimmed glasses he only wore in the mid-1980s. Seeing that shirt, I was transported back to the kitchen we had back then, with large floral wallpaper and an air-conditioning duct that smelled of Freon and mildew. There was even a tear in the elbow that I also remembered tugging on as a child, still present.

When I mentioned this memory to my mother, she gave me a look. "Anne, I bought your father that shirt just a few years ago." Sure enough, the label bore out her version of events – there was no way the shirt was decades old. My dad must have been hard on all his shirt elbows. Even worse, photo evidence reminded me that the kitchen I remembered didn't exist when I was a child. What I remembered was the kitchen of my teenage years – my childhood kitchen had blue plaid wallpaper. In the end, very little of my extremely vivid and personal memory was "true" or even possible.

I had confused and melded events just as Neil DeGrasse Tyson had done. My mistake was harmless, with the worst outcome being that I would have saved and treasured a shirt that was not the shirt I believed it to be. But false memories can ruin lives.

Satan, Toddlers, and Murder at Daycare?

About the same time I was standing in that plaid-wallpapered kitchen, Dan and Fran Keller were at the center of the 1980s "satanic daycare" scandal.[12] The stories children told of their experiences at the daycare run by the Kellers were nightmares: murders in broad daylight, with the Kellers cutting up bodies with chainsaws and burying them, sexual assault, children buried alive, and satanic rituals involving blood in the children's punch. These stories were brought out of the children by caring therapists, bent on getting the truth and bringing the perpetrators to justice. The most innocent-seeming details, such as the kids being given free American flags to take home, were steeped in evil once it was established the flags were an at-home reminder for the children to keep the daycare's secrets. In 1992, after testimony from children and a number of experts, the Kellers were sent to prison, where they would stay until 2013. Twenty years seemed like a light sentence for such horrendous crimes.

Except that there was no evidence of a crime. In an investigation and trial eerily similar to the Kellers', Ron and Linda Sterling ran a daycare in Martensville, Saskatchewan, Canada during those same years. They were accused of ritual abuse of the children, and the number of accusations rose as more and more families were contacted.[13] However, as reported by Lisa Bryn Rundle on the CBC radio program *Uncover*, "None of the children disclosed abuse until after police contacted their families. And it was only after 'repeated and highly suggestive questioning' that they made any allegations."[14] In one case she found that "Another child, who first denied that anyone had harmed him in any way, ended up saying after repeated interrogation by parents and police investigators that he had witnessed people being killed, acid being poured on people's faces, and that he was forced to eat feces and intestines, which were also stuffed into his ears." There was no corroborative evidence: no missing persons or dead bodies and no people with acid-burned faces. Yet these accusations and cases spread across the world.

In a 1992 report from the US Department of Justice, Ken Lanning summarized his findings from a multi-year investigation by the FBI:

> The idea that there are a few cunning, secretive individuals in positions of power somewhere in this country regularly killing a few people as part of some satanic ritual or ceremony and getting away with it is certainly within the realm of possibility. But the number of alleged cases began to grow and grow. We now have hundreds of victims alleging that thousands of offenders are abusing and even murdering tens of thousands of people as part of organized satanic cults, and there is little or no corroborative evidence.
>
> For at least eight years American law enforcement has been aggressively investigating the allegations of victims of ritual abuse. There is little or no evidence for the portion of their allegations that deals with large-scale baby breeding, human sacrifice, and organized satanic conspiracies. Now it is up to

> mental health professionals, not law enforcement, to explain why victims are alleging things that don't seem to have happened.[15]

Even more clearly, he stated on *Uncover*, "You can't murder twenty people in a house and not leave any evidence behind. I'll give you seven days to clean it up and you're not going to clean it all up. There's going to be trace evidence left behind."

As Lanning stated, mental health professionals were the explanation, though not the explainers, as Lanning had hoped. Therapists around the country were using techniques to "recover" traumatic memories in children and adults and finding an epidemic of graphic and ritualistic abuse. These recovered memories appeared even for people who had always thought their childhoods were safe and even pleasant, once they fully invested in the memory therapy. This therapy was achieved with a few techniques. One was called "guided imagery," a method of accidentally implanting false memories that was, shall I say, devilishly effective.

Guided imagery works by changing memories when the noodle is in a soft state, as bits of memory are recalled. In my own example, parts of the kitchen were present in my memory and in actuality – the smell and location of the floor duct. Guided imagery starts the client off with those bits of memory, and then the therapist encourages the client to remember other details of the kitchen, helping to build a vivid memory world that may or may not have existed. Beth Rutherford described her experience with guided imagery on the radio show *This American Life* in 2002. She was convinced by a therapist that she had been abused by her father as a child. Beth explains the technique and its consequences like this:

> ... OK, now, think about where you were as a child. OK, think about sitting on your bed. So I'm thinking, thinking. OK, now what are you wearing? So I'm thinking, thinking. What's an outfit I can remember wearing as a child? And then you just build on that, and build on it. And what did that feel like. You know, it's kind of like you have a little half teaspoon of memory, so to speak. That little half teaspoon turns into a whole three tiered cake ...
>
> And then I was given books to read about stories of people that had been abused. And she'd actually pick out certain

> chapters for me to read. And it began obviously to preoccupy a lot of my time, thinking about this. Trying so hard to remember if something had happened. And I couldn't remember, and I'd try to think even more …
>
> I remember the first time it happened, it scared me to death, because I kind of looked up to her. And she said to me, do you know what just happened? I said, no, what? And she said, you just recounted for me a story of what had happened to … What your dad had done. And she has a piece of paper. And she's told me she wrote down everything I said. And then she started, and she began to read back to me. I can hardly describe the horror, to sit there and listen to that. She's saying I said it. She's reading it to me, as I described an event where my dad had brought me to my mom and his bedroom and laid down next to me. She's describing, and it was so horrifying. I can't hardly describe for somebody what it's like to believe that you have been loved as a child, and you grew up in a wonderful home. And to sit there and listen. And it's literally like the foundation of your life is coming apart.[16]

Eventually all of the imagery the person is absorbing, through books or imagination, is integrated into those vivid memories of childhood, until they are inextricably linked and feel as real as we expect an actual memory to feel. This was a situation where blame is difficult and often misplaced. At the time, guided imagery was an accepted (though not evidence-based) therapy, and the therapists who practiced it believed they were helping their clients uncover true events and bringing evil-doers to justice. For those undergoing the therapy, there was *no physical or neurological* difference between what we consider "real" memories and these implanted false memories. The therapy made them believe and *remember* that the false events actually happened. Everyone was acting in good faith, even while previously happy families were split apart and innocent people went to prison. It is worth noting that the problems with guided imagery were well known by the 2002 broadcast of *This American Life*, yet the Kellers would remain in prison for eleven years more.

Believe the Children! (When Interviewed Correctly)

Finally, memory is most malleable in children, making it critical that when they provide evidence, they need to be interviewed by experts trained in asking non-leading questions. These interviews should be recorded – not only for the children's answers but to provide a record of the questions. There are numerous scientific studies showing the need for these measures, with one led by psychology professor Debra Poole providing a vivid and memorable example. Poole and colleagues held a Science Camp for kids aged 3 to 7.[17] At this camp, "Mr. Science" led the kids through typical age-appropriate science experiments: baking soda and vinegar mixtures, reaching for moving objects while wearing prism glasses, and squishing silly putty onto a newspaper to lift the letters. When interviewed right afterward, the kids were accurate in saying what activities they did and what the camp was like. But the researchers had another activity planned, and bided their time.

About three months later, the kids' parents received a "report" of what went on at the camp. In this report, two of the science activities were described accurately. But two of them were entirely made up. All were read to the kids by their parents (the parents did not know that some of the activities were fabricated). In addition, the report falsely stated that at the end of Science Camp, Mr. Science wiped each of the kid's faces with a wet wipe that "tasted yucky."

The kids were then asked to tell the story of what went on at Science Camp. Almost half reported that they did one of the made-up science experiments. When kids were asked leading questions like "Did you make a paper airplane with Mr. Science?" (spoiler alert, they did not), almost 95 percent of the children said they remembered at least one of the made-up science experiments. Over 70 percent falsely remembered that Mr. Science put something "yucky" in their mouths.

If these results scare you, you're in the same boat as I am. Children are clearly highly susceptible to having false memories implanted, which can cause them great harm later on as well as to the adults in their lives. Thankfully, researchers have discovered how to interview children. The US Department of Justice created and disseminated some best practices found in the research literature for interviewing kids 3 and older, such as to record the interview electronically with both audio and video, use open-ended questions throughout, delaying focused questions as long as possible, allow the child to describe a recent non-abuse experience in detail as practice and to build rapport, and to ask the child to describe his or her abuse experience in detail, and not to interrupt the child during this initial description.[18]

What should scare us all more is that, although there are evidenced-based best practices for interviewing children (and adults) about their memory of events, there is little, if any, regulation. Media has brought some of these issues more into the public eye, such as the recorded interview of Brendan Dassey in the Netflix documentary *Making a Murderer*, but jurisdictions are independent and it will take the public demanding regulations informed by science to see an improvement.

Despite the fallibility and malleability of memory, none of the above applies to the memory of survivors of trauma. This is a contentious issue, as those survivors rightly believe that defense lawyers will try to use the research findings on false memories to call into question their very real memories of abuse and attacks. Though we all have false memories, unless we are manipulated with leading questions or guided imagery, the *gist* of our memory is true. Though the details may be proven incorrect or even change, that does not affect the truth of the main memory. In my own harmless example, the image of my father standing in the kitchen ready to go hunting was a true one, even if the exact shirt he was wearing differed. For survivors of rape or abuse, the memory of the act is true – even if the exact shirt the attacker was wearing differed. False memory research should not be used to question the veracity of those well-remembered events.

The science on techniques for "recovering memories" is fairly clear: competent therapists should avoid them. The American Psychological Association (APA) states that

> First, it's important to state that there is a consensus among memory researchers and clinicians that most people who were sexually abused as children remember all or part of what happened to them although they may not fully understand or disclose it. Concerning the issue of a recovered versus a pseudomemory, like many questions in science, the final answer is yet to be known. But most leaders in the field agree that although it is a rare occurrence, a memory of early childhood abuse that has been forgotten can be remembered later. However, these leaders also agree that it is possible to construct convincing pseudomemories for events that never occurred.[19]

In choosing a therapist, the APA recommends,

> First, know that there is no single set of symptoms which automatically indicates that a person was a victim of childhood

abuse. There have been media reports of therapists who state that people (particularly women) with a particular set of problems or symptoms must have been victims of childhood sexual abuse. There is no scientific evidence that supports this conclusion. Second, all questions concerning possible recovered memories of childhood abuse should be considered from an unbiased position. A therapist should not approach recovered memories with the preconceived notion that abuse must have happened or that abuse could not possibly have happened. Third, when considering current problems, be wary of those therapists who offer an instant childhood abuse explanation, and those who dismiss claims or reports of sexual abuse without any exploration.[20]

A "Good Memory"

The theme of being all too human, with our fallibilities and limitations, occurs again and again. I highlight failure, because failure is what we must fight to overcome through the design of our systems and the technological world around us. But when I talk about memory, even with the evidence of our own lapses all around us, I inevitably get a "what about" response from my students, my friends, and even my family. After all, our major memory failures make the news (such as with the scuba incidents), but examples of incredible feats of memory are also out there. "I don't have a good memory, but what about the guy who remembered all those digits of pi?" I looked him up, and indeed Akira Haraguchi was able to quietly sit with his eyes closed and repeat back 100,000 digits of pi over sixteen hours.[21] Certainly more than I can remember, even using the cheer I learned at Georgia Tech: "Cosine, Secant, Tangent, Sine... 3.14159! Go Jackets!" Or what about competitors at the World Memory Championships, who can glance through decks of cards, put them aside, and repeat back all the cards, in order? Surely these memory masters would never leave a sponge inside of a patient or forget to go running almost five years into a running streak?

I'll counter these claims of superhumans by saying that no one has "a good memory." We all have limits, with some variation. Sure, our memories for new knowledge tend to be better and more flexible when we're in our early twenties than in our eighties, but that differs by person and the differences aren't that big in the grand scheme of things. It is the same with other types of memory, such as our working memory, where we have differences between people, but those differences are not large – certainly not enough to encompass 100,000 digits of pi. So how did Haraguchi do it?

The answer is that we mistakenly think of memory as an ability, a capacity that you are born with or comes naturally to you. Memory, however, is a skill that can be gained, but that skill will always be specific to what was practiced. Haraguchi may be able to recite pi for sixteen hours, but I don't expect him to remember his grocery list any better than another man his age. Unless, of course, he has also practiced remembering grocery lists in his spare time! I don't expect that the card memorizers would be more likely to remember my name and face, or really, remember anything else besides card names and card order. But I don't want to downplay the amazing *skill* Haraguchi showed in his recitation. Such skill takes time and dedication far beyond what most of us are willing to commit. But it is fascinating to see how he did it and to know that any of us could do the same, given time, opportunity, and practice.

Alex Bellos interviewed Haraguchi for *The Guardian* in 2015 to ask him how he was able to remember so many numbers in a row. Haraguchi said,

> I have created about 800 stories, whose lead characters are mostly animals and plants. For the first 100 digits of pi, I have crafted a story about humans. Here is how the first 50 digits, starting with 3.14, reads: "Well, I, that fragile being who left my hometown to find a peace of mind, is going to die in the dark corners; it's easy to die, but I stay positive."[22]

It's easy to die, but I stay positive. Poetic if depressing, his example showcases the method of developing a skilled memory for ordered numbers: turn the numbers into meaningful words and strong images. This is the same method used by the card memorizers and by almost anyone able to demonstrate a seemingly incredible memory. If you'd like to try it, follow these simple steps:

1. Collect all the items to be remembered. For Haraguchi, that starts with the digits 0–9 but likely evolves into longer strings of digits

(0–999). For the card memorizers, it's more constrained: 52 cards and then eventually pairs of cards (for example, if the Queen of Clubs follows the Two of Spades, it exists as one item. Do this for all combinations.)

2. Give each item a visual representation. For example, I might decide that the Queen of Hearts is Beyoncé, and practice thinking of and seeing an image of Beyoncé every time I see that card. For numbers, you can do the same (I think of my niece Savi every time I see a 6, for some reason) or you can use a method where the sound of the number signals a word (four sounds like "rrrr" and so one could visualize a railroad). The second method is what Haraguchi used, though with Japanese language sounds.
3. Practice for many, many, many hours associating the visuals you chose with the items. Use flashcards, slideshows, and phone apps. Practice until seeing one of those items makes you visualize with no effort – Beyoncé just "pops" into mind. This will take a while and a lot of dedication, but it will happen if you keep working at it. You can thank the association cortex in your brain for this.
4. In the final step, practice making stories with your visuals. One popular method is called the "memory palace," where you think of a building you know well (or build one in your imagination beforehand) and mentally wander through the building as you memorize. Turn over the first card, it's the King of Clubs. "Ah, Richard the Third!," you think, and immediately picture that club-footed monarch in a funny hat kicking at the door. Next card is the Three of Hearts, your childhood dog Sparky, who struggles to open the door with her paws, barks, and backs Richard III into a corner of the entrance foyer. Next card is the Nine of Spades, which you memorized as your friend Brian (which sounded to you like Niiinne – a stretch, but it worked). "Sparky, get down!," shouts Brian from his seat on the sofa, moving you into the living room. And so the story continues, with each new character interacting with the last, as you move through the house.

Congratulations. Now you have "a good memory." Unfortunately, it's only for numbers or cards, so you're still going to forget that dentist appointment unless you put a reminder in your calendar.

Fortunately, reminders work. Language and abstraction, two of our strongest human abilities, helped us develop supports for our relatively weak memory skills. Our strategies are as numerous and creative as we are, ranging from putting important items by the door, so we can't miss them when we leave, to asking a smart speaker to verbally tell us what to do on a certain day and time. It's when our daily tasks become sufficiently complex that these reminders are crucial. LeClair and the thousands of other patients with surgical instruments left inside of them deserve our best efforts at inventing and regulating systems that aid healthcare workers in their attention and memory.

Beyond the Checklist

Instructions regarding surgical instruments can be on a checklist, such as requiring a baseline and final count of the hundreds of implements used in a typical surgery. But even with these steps in place, foreign objects still get left inside patients. However, by carefully examining everything surgeons and support staff have to do, and armed with our knowledge of human capabilities and limitations, it is at least predictable when simple checklists could fail. We can use the same information to design stronger tools than the checklist alone, and we can use it in areas outside of surgery.

A surgery that leaves a patient with a retained object tends to have particular qualities.[23] First, the surgeries tend to be emergencies or to have unexpected changes occurring during the procedure. Second, the surgery location is usually the abdomen. Our torsos are large places, with many nooks and crannies for a sponge to get lost in. Compare the opportunity for lost objects in the abdomen to a wrist surgery to correct carpal tunnel syndrome and it's no surprise the belly is on top. Third, 70 percent of items left behind are sponges and gauze. There is the occasional plastic tube or surgical clamp left in place, but the soft and absorbent items make up the large majority. This might be expected because these soft items take on the color and apparent texture of the inside of the body, effectively camouflaging them. Also, the soft items are also the most numerous with the possibility of counts in the hundreds for an abdominal procedure. They are used for absorbing fluids but also for staging and separating the target organ from others nearby. Complicating this further is that they may be cut into pieces for a good fit (though this is discouraged). Lastly, some procedures require sponges to be packed into the wound and remain there when the patient moves to recovery. The baseline count has to be updated, often in a fast-moving and stressful environment.

With all of these risk factors, one might start to be amazed that items aren't left behind in every patient. The relative rarity of retained items is not due to careful healthcare workers – it is due to the systems instituted in hospitals to support the memory and attention of those workers. Some hospitals use high-tech scanning systems, where every surgical tool, including sponges, has a barcode that must be scanned upon opening and then again at the close of the procedure. The Mayo Clinic was a shining early example of how scanning tools improved outcomes, moving from an average of one object left in a patient every sixteen days to *not a single object left behind for eighteen months.*[24] Doctors and nurses everywhere *want* to remember all surgical items and greatly fear that they might leave something inside a patient. But it is impossible for a person or a team to remember everything – so we must design systems that prevent otherwise-guaranteed memory failures. That goes for scuba trips and running streaks, too.

Contact Tracing

The COVID-19 pandemic gives us another opportunity to save lives through science. I'm not talking about vaccines or better hospital treatments, although there is a psychological element to those as well. I'm talking about contact tracing.

Tracing sounds simple enough. Just ask someone who has the novel coronavirus whom they've been near in the last two weeks. But we know enough about memory to know it would be easy to miss a contact, especially with the stress of just learning that one has the virus or was recently exposed. This is why the Centers for Disease Control (CDC) have created training to guide contact-tracing interviews. I went through the training myself and was pleased to see how research in human memory was translated into clear advice. For example, here is a sample script for an interviewer where I've italicized the cues and anchors that should help prompt the most accurate memories:

> I am going to ask you to think back *over each day* while you have been sick to remember *what you did each day*. This will help us figure out who you may have been around, and who else might get sick. If you are having a hard time remembering, sometimes it is helpful to *look back at a calendar*, or on *your phone for messages* sent on each day, or even at your *credit card*

> *receipts*. Please list all *activities*, *places* visited, and *travel* you participated in starting [48 hours before the day of their first symptom as calculated by the interviewer].[25]

Below are prompts the CDC provides to the interviewers:

- Where did you wake up *on that morning*?
- Did you go to work or school on that day?
- What is your work or school environment like?
- What is your *normal* work or school day like?
- Who lives with you?
- Did you have any *visitors*?
- Who did you eat your *meals* with?
- Did you have any *outings* or *social gatherings*?
- Did you ride on *public transportation* or in a *ride-share*?
- Did you have any *appointments*?[26]

All of these questions can prompt other memories, and the use of "contemporaneous evidence" like calendars and receipts are also excellent prompts to remember a specific contact.

Conclusion

The protagonist in Barbara Kingsolver's book *Animal dreams* searches for truth in her childhood memories, though often faced with evidence against them or that she has forgotten events well remembered by others. In many cases, she remembers the gist of a memory, but the details could not have occurred. She sums up this experience by calling memory "a relative to truth, but not its twin." Through fiction, Kingsolver captured truth about the experience of being human. We forget, we forget that we forgot, we remember events that never happened, and we remember things that did happen. In most of those cases we have little control and no way to tell a 'true' memory from a false one. There are many existential threats to our memory, from Alzheimer's disease to dementia and amnesia. But even at our best, with no pathologies at play, our memories are unreliable. We can train our memories, but experience little transfer of that skill to areas beyond our training. For creatures who exist as the sum of our memories, this is a strange place to be.

It is our adaptability that saves us, that makes us human. We create tools to hold our memories, ways to stop our failures. The written word carries the memories of people who died long ago, transcribed for us and those who come after us. We make calendars, checklists, and mnemonics. We build systems and sensors to catch our memory failures before or as they happen, allowing us to achieve incredible successes despite our poor memories. I would like to leave you with that thought – that memory issues are just one more way that we are all too human, but that we can control how we build in fail-safes to reduce the harm. Memory "errors," from leaving divers in the water to retained surgical instruments to unreliable eyewitness testimony to keeping a running streak going, are all preventable with a good system in place.

11 OLD PRINCIPLES FOR NEW WORLDS

> We will control the horizontal. We will control the vertical. We can roll the image, make it flutter. We can change the focus to a soft blur, or sharpen it to crystal clarity. For the next hour, sit quietly and we will control all that you see and hear.
>
> – *The Outer Limits*, TV show, 1963

I have an intimate familiarity with the dangers of virtual life.

In 2014 I was invited to review grant proposals for the National Science Foundation (NSF). These panels are a serious time commitment to help the NSF to decide how to award millions of dollars in research funding each year. I treasured the opportunity to serve, as I enjoyed reading the proposals scientists submitted and the deep and wide-ranging discussions with other researchers about the merits of each project. Reviewers travelled for two days of meetings in Washington, DC.

This time there was a different plan. This panel was going to save money and time by meeting in a virtual world: *Second Life*. Linden Labs launched *Second Life* in 2003 as a free-to-access virtual world. By 2014, it had around a million users socializing with each other while making their own goods, currency, services, and fashion. The NSF had an island in *Second Life* where we could meet privately in a large space, see each other as 3-D avatars, and discuss the proposals over our microphones and speakers. Because I live in North Carolina, travelling to DC was never difficult for me, but it was much more of a commitment for scientists in the western USA. I'm sure they welcomed the option of *Second Life*.

I downloaded the program a few days prior to test my microphone and work out the kinks. I was not inexperienced in virtual worlds or how to move about in them, having spent many years playing first-person video games. The interface in *Second Life* was unfamiliar and complex, but I managed to cobble together an avatar that looked somewhat like me, with brown hair and a professional outfit. The microphone seemed to work, so I was set and spent the next twelve hours making sure I knew the proposals well enough to discuss them.

On the morning of the panel I entered the *Second Life* world and was immediately greeted by strangers who had no concept of personal space. They crowded in on me and my view was blocked by their polygons and gesticulations. Fortunately, I could take off my headphones so I didn't have to listen to them. I looked back at the instructions from NSF and managed to teleport out of the melee to the island – a quiet reprieve surrounded by water with just a few of us under an open pavilion about the size of a gymnasium.

Inside the pavilion was a conference table surrounded by stools. Floating on the outskirts were two large screens, where we could view the ratings and status of each proposal as we discussed. Eight respected scientists from around the country sat on the stools, dressed in different levels of reality and formality. One man had blue hair, another wore a Hawaiian shirt. The others wore suits, excepting one woman in jeans and a sensible charcoal t-shirt befitting a scientist. Leading us was NSF Program Officer Bill Bainbridge, a 73-year-old sociologist and expert in virtual worlds. He knew all the tricks, conjuring up objects for us like a magician while laying down the rules for discussion. We got down to business.

The pros and cons of a virtual meeting quickly became apparent. Taking turns talking was as trying as any analog conference call, filled with interruptions and "oh you go" and "no, please, go ahead." In person, I could tell who wanted to speak next, because they expectantly leaned forward at the table with their neck stretched out and slightly raised eyebrows. Our avatars didn't do that. We also lacked the coffee and pastries usually on the table at the back of the room.

But there were benefits to the virtual world. First, I am terrible with names and worse with faces. This was always a challenge at in-person panels, because after brief introductions, no one ever repeated their name. But in *Second Life*, everyone's name floated gently around their head. No memorization necessary. Second, when someone was talking, a little bubble appeared over their avatar. This was much better than a conference call, because I saw the name of the speaker and the speech indicator and knew who was talking. It made it quickly evident who was the blowhard who wanted to monopolize the discussion (and was deftly intercepted and redirected by Bainbridge).

After heated discussion of three proposals, talk moved to a proposal not assigned to me. I was interested in it, but admit that my mind wandered. 'I think my avatar would look smarter with glasses,' I mused and pressed the key to bring up aesthetic options. A pair of black horn-rimmed ones looked good and I clicked on what I thought would add it to my character.

Imagine my surprise when instead, my avatar stood up from the table and dropped all of her clothes on to the floor.

"Oh, shit!" I screamed, fortunately muted. I frantically pressed keys and clicked on menus, but my avatar had decided to free herself of the chains of decency and stood, facing away from me and toward the group. I did the only thing I could think of, which was to close *Second Life* entirely. I would have pulled the plug off the wall if I had to.

Sitting there with my pulse slowing I considered my options. If I joined again, would I still be naked? Would I still be on the island? Suddenly the crowds of strangers didn't seem so off-putting, if it would give me time and a chance to figure out how to put my clothes back on before reentering the professional space. I reloaded *Second Life*.

When I entered the game I was back on the NSF Island and clothed. What a relief. We continued the discussion. No one said anything. To this day, no one has said anything. My only hope is that my colleagues were so consumed by the discussion and their own appearance that no one noticed mine. My avatar still wasn't wearing glasses.

Looking back, most of the causes of my embarrassment were due to the discrepancy between my expectation of how the world works and the choices of the system designers. In the physical world I can try on a pair of glasses without taking off all my clothes (much appreciated by the staff at my eye doctor). I assumed that would also be true in *Second Life*. The error was exacerbated by an interface that didn't offer clear or fast options to return to a state just previous to the error such as "cancel" or "undo." I wish I could say that *Second Life* was the only form of technology with such a mismatch of human expectation and design decisions, but similar issues are rampant in today's technology. The good news is that we can do something about it.

Every detail of virtual or computer-mediated communication was designed by people. The system may feel fait accompli, but every choice made within a design can be changed and improved. There are well-known methods for improvement and they all depend on these core principles: understand the people for whom you are designing, and understand what it is they must do to accomplish their goals. Human factors experts usually shorten this to say, "Know the user and know the task."

Virtual worlds are a microcosm of each of the human strengths and weaknesses discussed so far. In virtual worlds we experience the limits of vision, attention, memory, the pros and cons of our social nature, the impact of diversity on human creativity and problem-solving, and predictable biases we carry with us throughout our lives. In worlds entirely dependent on technology, from videoconferences to virtual reality, our humanness shows up all the more and must be considered in the design of those worlds.

Mental Models

Many parts of our lives, from social interactions, to work, and even household chores are mediated by technology. This is not new. The first time humans chiseled words into wood or stone for another person to decipher in a different time and place was a technology. Books, letters, phonographs, daguerreotypes, Morse code, e-mails, text messages, multiplayer video games, and videoconferences – all are liaisons for human communication. Often the thoughts, emotions, and ideas they spark are identical to speaking face to face, but in many cases the format constrains intent and communication. Communication issues can be inherent to the medium or it can come from problems with the design of the system. Design problems usually come from a lack of understanding of human capabilities, and limitations or lack of understanding of what people are trying to accomplish with the technology.

My birthday suit appearance in *Second Life* showcased that the designers probably didn't consider that novice users would be forced into using their world because of work. They probably assumed any player chose to be there, for fun. But I and many others were reluctant users with little time to learn. For the 'task,' meeting with a group of strangers to discuss and record the merits of scientific proposals, the system was flexible enough to allow it. But it wasn't built for the etiquette required in those situations. In a more typical *Second Life* interaction, momentary lack of clothing might be commonplace or even desirable. But it was not desired during the serious and professional endeavors of a grant review panel. My point is not that *Second Life* should have anticipated this niche use of their product. It is only that we *can* anticipate such uses and design for them. If we want an easy-to-use and efficient world around us we need to demand that technologies are properly designed for us and for our goals.

The Toolbox

The tools that human factors and usability professionals use to understand their users and their users' tasks include interviews, surveys, and observations. These are created and considered with a generalized understanding of human merits and shortcomings. It's safe to assume that *every* human has limited working memory, and thus can't hold a dozen items in mind when listening to an audio menu. But to know what menu items are most important, or where submenu items should fall, requires a specialized understanding of the people who are going to use that system. An accountant using tax software is a far cry from me fumbling through TurboTax, even if we have the same memory ability. That's where the interviews, surveys, and observations come in.

Moving on to the task, or the job that is going to be performed with the system, researchers use the formal method of *task analysis* to break it down into steps. How do people get to, or try to get to, the outcomes they want? A task analysis records each step. Then each step can be examined and possibly changed. It's hard to overstate the impact this simple-sounding process has had on the science of human behavior. Task analysis helped designers decide which steps should be done by people and which should be offloaded to machines, as in the chapter that covered the "Miracle on the Hudson" (Chapter 1). Task analysis could have found the error that allowed the river to taint the water for the people of Flint (Chapter 2). Task analyses can give insight into both the mind and behavior.

There are many forms of task analysis, from a behavioral task analysis that steps through the task in chronological order to cognitive task analysis, which examines in depth the knowledge, skills, and decision-making that people bring to a task.[1] Though these task analyses can follow formal rules, they can also be flexible enough to focus on the most important questions. For a new surgical device, each step might be coded for how disastrous an error would be, but for a video game each step might be coded for the likelihood it frustrates or misleads the player. Steps in the analysis might be purely physical, like "add anti-corrosion chemicals as water treatment," or they might focus on the required human decisions and whether the task-doer has enough knowledge, skill, or feedback for each decision to be the right one. This might involve asking the person about previous outcomes with questions like, "What are some examples when you have improvised on this task, or noticed an opportunity to do something more quickly or better, and then followed up on it?"[2]

Users and Tasks

Getting to know the people who will be using a technology can be tricky. We all have different roles in life and the same person can act differently at home with family than they do at work or out with friends. Maybe I'd love attending a virtual nudist colony on the weekends, but it's a different story when I'm at my job and getting paid to be there. (On second thought . . .) Information about the person needs to be considered in the context in which they will be using the system. The key to knowing who they are and what they want is listening, asking open questions, understanding what they already know, collecting their likes and frustrations, all the while grounded in an understanding of where humans excel and where they tend to fail.

Perhaps even more key is entering into this data collection knowing and truly *believing* that you don't already know who users are or what they want. Designers must eliminate their assumptions, but that's hard to do. All of the biases in previous chapters, from confirmation bias to irrational decision-making, are our all-too-human qualities that push us to make assumptions. This is why we need rules to follow in knowing the user and knowing the task.

When a process is formalized with rules and constraints, it becomes a tool. Fortunately, we humans are great at inventing tools. On the surface, task analysis seems simple. Record all the actions the person needs to take to succeed. Design a system that does those actions. But a formal task analysis does so much more than that, and is so flexible that it is the core of every good design and improvement of technology. A formal task analysis takes a goal and breaks it down into subgoals and all of the steps needed to achieve those goals. It can go as deep as it needs to go and cover the crucial needs of the users. The deficiencies in most videoconferencing software illustrate the importance of a thorough task analysis.

Zooming from Meeting to Meeting

Videoconferences were once solely a domain of industry, most often carried out in conference rooms inside tall towers made of steel and glass. But as large numbers of employees started to work from home, key attributes of the users and the videoconference changed. Videoconferencing became necessary to teach classrooms of small children and to tutor teenagers in the subjects they were missing due to the COVID-19 pandemic. The "social" videoconference became an event, adding more screen interaction at the end of the workday. The brand *Zoom* became as synonymous with meeting as Kleenex was with tissue. But even workers experienced with videoconferencing faced new challenges when working entirely from home: they had children and pets to care for, interruptions never present in the office, and no chance to gather socially and build trust (or distrust) of each other. Talking heads couldn't convey all of the body language cues present in a physical space. All of these changes and adaptations called for design and feature changes in videoconferencing to enable *these* users to achieve *their* goals.

Instead, videoconferencing took a mental and physical toll on its users. In previous chapters I covered how limited human attention is and how it's divided into resource types (Chapters 4 and 5) . In videoconferencing, it's easy to see where the attention of the host, and probably the conference participants, will be overwhelmed. First, almost all of the information from videoconference software is visual. Some allow participants to "raise a hand" that appears next to their name. Near their names is other information: Are they muted? Are they speaking? Are they connected via a computer or a phone? If chat is enabled, text may be appearing and changing as more is added. If the chat window is hidden, a notification flashes when a person types new text, attracting everyone's attention. Some systems have a waiting room, where each person joining must be cleared for entry by the host. (And don't forget to check for latecomers.) If breakout or side rooms are used, people must be assigned to them by the host. The host also needs to monitor their own microphone and camera, muting if there is background noise or if others need to be heard clearly, while perhaps monitoring everyone else's mic (considering how commonplace it is to forget to mute when not talking). As the memory chapter (Chapter 10) made clear, people *will* forget to mute and blaming them for it won't solve the problem.

Every one of these alerts, decisions, and actions requires visual-spatial resources and most require a great deal of visual-verbal resources (to read words). In the chapter on recognizing objects (Chapter 9), I discussed how human eyes evolved to detect motion in their periphery, so good luck trying to ignore blinking notifications, even if they are irrelevant. The sheer number of things that must be attended to overloads the host's controlled processor, and the fact that most indicators are visual means that the attentional 'fuel' is depleted from only one attentional resource. It is no wonder that hosts of large meetings, whether it be for business or a classroom, struggle to keep track of the interface even as they attempt their main job of presenting, teaching, or managing others. Every time I hear a meeting host apologize I want to say "It's not your fault! The system is against you!"

Success may be on the horizon thanks to engineering breakthroughs and a focus on human needs. The graphics card maker NVIDIA released a kit that can be used by videoconference companies to improve the social interactions they support. It has a number of impressive features, and I have categorized them into whether they focused on making videoconferencing easier and more pleasant by 'knowing the user' or 'knowing the task.'

Knowing the User

Eye contact and face orientation: For a person to make eye contact with someone in a videoconference, they have to look away from the face and into the camera. This gives their partner eye contact, but leaves one person staring into the HAL-like lens of a camera. In this zero sum game one person is always left without eye contact. The NVIDIA system uses computer graphics to reorient the users face and eyes so that they appear to look directly into the camera, even if their actual face is turned away or their eyes are looking down at another face on the screen.[3] Thus, this feature returns one cue of social interaction to the desolate landscape of the videoconference.

Knowing the Task

De-noise: Background noise is hard to control when working from home, and the NVIDIA system acknowledges that workers might not get to make calls from the peace of a conference room or a closed office. The new system eliminated any noise outside of the speaker in front of the microphone. The video demonstration showed the loud cries and xylophone bangs of two toddlers reduced to silence, even though they were a mere 8 feet from their mom and the computer. The speaker still has to suffer the distraction of background noise in their own home, but at least the other attendees don't have to suffer along with them.

This will be an interesting feature to explore further, because it may not be only a benefit. It may be that the distracted speaker, stuck in the loud chaos of their own environment, comes across as distracted and unprofessional. Before de-noise, the others in the videoconference would know why the speaker was distracted, and perhaps have some sympathy. But with de-noise it may eliminate the obvious reasons for strange behavior on the part of the speaker.

Virtual backgrounds: The visual twin of 'de-noise,' a virtual background replaces the environment behind the speaker with a photo or video. This removes distractions or parts of the user's life that they wish to remain hidden, such as clutter or a bed. Such backgrounds are an acknowledgement that many workers aren't able to dedicate space in their house to a professional office set-up. Potential downsides include the same issues as de-noise: if other people or distractions are present in the room with the speaker, those distractions will affect them but be unknown to those watching the videoconference.

Conversational avatars: Videoconferencing has expanded beyond work into fun and whimsy. A conversational avatar is a mask for the user's face, but one that is fully actuated with face movement and expression. It is a more Hollywood-ready version of the Animoji (the moving version of Memojis) that Apple created. The visuals are only limited by what the artists provide, allowing a talking head to appear as anything from an alien to a dragon. These avatars could be used for anonymous interactions or for playing remote games that require conversation.

All of these improvements in videoconferencing are general and likely to apply across multiple objectives, from work to online family get-togethers. However, some jobs have very specific demands that aren't met by video conferencing, such as music lessons. Again, understanding the users and their tasks led to a better design.

Maribeth Gandy Coleman is the Director of the Interactive Media Technology Center at Georgia Tech and an expert in mixed-reality technology. She's also an avid runner and an efficient multi-tasker: we talked for about an hour about technology-mediated social worlds as she steamed along behind her new puppy, Kylo.

She told me about a student research project she supervised in the spring of 2020. The student discovered that music teachers, forced to go online, were having a hard time with lessons. Videoconferences were optimized in two ways: to accurately transmit the human voice and to do so using as little internet bandwidth as possible. However, this meant that the sounds of a flute being played were cut off. "Some of the frequencies in the signal that created the critical nuance in a skilled flautist's performance were being removed," said Gandy Coleman. "The tool my student built actually mapped that lost frequency information to a visualization so the teacher could listen but also see the rich details from the performance in real time." This was an engineering solution, developed in an audio engineering course, but it was based on the human-centered design principles of knowing the user and knowing the task.

Mental Models in Professional Virtual Worlds

With a good background in the importance of designing technology for communication to match the needs of and qualities of the humans using it, let's return to communicating in a virtual world with all of its potential benefits and pitfalls. As with videoconferences, VR conferences offer a chance to apply all we know about human limits and capabilities in design.

To learn what it was like to attend a conference in virtual reality, I spoke with Laura Levy, a research scientist at the Interactive Media Technology Center at Georgia Tech, about some observations she collected at a virtual reality conference hosted by the IEEE (pronounced "eye triple E"). An avid gamer, Levy was sent there to observe the struggles and successes of the conference attendees. With hundreds of attendees and all personal interactions mediated entirely through technology, it was an expansive microcosm for how all of the human capabilities and limitations discussed in this book affected life in a virtual world.

As attendees were mostly computer scientists and VR enthusiasts, the conference was set up to succeed. Many attendees owned and used VR headsets, allowing them a 360-degree view of the conference world. Others used computers, but were able to move around with a first-person view. Areas were spatially laid out, with universities creating models that mimicked the buildings on their campuses, along with more imaginative spaces, like a snow-covered night landscape peppered with fire pits. Social interactions were built to copy real-world interactions, such as going up to someone to speak with them, entering a room where a speaker was giving a talk and listening to that talk, being able to ask questions after, and even social events in the evenings where researchers with similar interests could gather for networking and discussions. But before I paint too rosy an experience for the VR attendees, it's worth discussing the mismatch between human knowledge and the manufactured reality of the conference.

Rooms Full of Dead Puppets

When I'm at a conference I get distracted. I need breaks. Maybe I need to send a text, so I run off to a corner of the hotel and hide behind a plant, trying to get away from the bustling conversation and elevator music. Or I duck into a bathroom for a short respite and send an email. Attendees at a virtual conference need these same breaks, maybe even more if the dog or baby needs attention in the next room. The virtual world of the conference was designed so that, when a person stopped actively controlling their avatar with a VR headset, the avatar stayed in place, head down and arms limp. Walking around the conference meant numbers of these driverless puppets standing despondently in hallways and atriums. Practically, they were in the way. Psychologically they suggested just how many attendees were mentally checked out.

Even when the puppets were active, it could be hard to talk with each other. Humans have mental models created from years, even decades, of experience that help us predict what is likely to happen. In social situations, these models are usually accurate. The next time you're in a group, notice the little nonverbal signals each person gives that they want to speak next. It should be clear to the group, boors excepted, who is next. A change in posture, raised eyebrows or chin, lips parting – all are signals we are unconsciously attuned to. Our automatic processor files away these cues and helps us take our turn and interact in a comfortable and effortless manner. Unfortunately, in the virtual conference, these cues were erased. Attendees were left with awkward interruptions, made more awkward by the design of audio systems that cut in and out depending on who was speaking. The automatic parts of conversation became those that needed constant effort and attention. Attendees were exhausted.

Rooms Full of Hot Mics

Laura Levy told me another story about conversations in VR that made me cringe, though it did illustrate how much we depend on mental models of how the world works.

> In a real conference, you can whisper to somebody next to you, and it's not that big of a deal. People were trying to whisper to one another, as they were watching the presentations, but didn't realize that most of the rooms didn't have any attenuation of audio. So, if you were speaking at one volume all the way across the room, people heard you just that loud on the other side. They didn't realize that they were speaking at full volume while someone was presenting. Now, the presenter didn't know anyone was speaking because they were on a completely different platform, not in the "room" we were in. The other people in the audience didn't really know what to do. You could see people turning their avatars around trying to figure out who was talking. Someone said "Hey, you know, we can hear you" but since the other person was talking the sound didn't get through to them. I heard some intimate conversations about emotional abuse and internal politics of businesses. I don't think these people would be okay if they knew that the whole room was hearing that, but they didn't know. It was awkward. But do you go up and tell them, "This whole conversation you've had how you hate your boss? Everybody heard it. And also, your boss might be here."[4]

The thought of strangers and co-workers listening in on what I whisper to a friend during a talk was chilling. It can be dangerous and disruptive when technology violates long-held mental models. The attendees eventually made discoveries about their world and new tactics emerged. They found that in some rooms, if you moved away vertically (flying was allowed), the sound was attenuated by distance and you could have a private conversation. When done, they would rejoin their groups by flying back down from the ceiling.

It was fortunate that the platform used by the conference was flexible enough for adaptation. It was fortunate that the conference attendees were people with the technology skills and experience most likely to adapt. But keep in mind that the entire experience was designed by humans, including using a mental model of sound more appropriate to a game where far-flung players need to communicate. Instead of depending on human adaptation and willingness to conform to the system, imagine a system designed to work with people in their goals and created to support their capabilities.

A Conference without a Birdfeeder

What Levy found most lacking in the VR conference were coffee breaks, and not just because she lives for coffee.

> People really missed the social aspects of the conference, even though there were social rooms where they could go. But what was missing for them was the coffee. Well, not the coffee itself. They don't really want the coffee, but they want a reason to be gathering around this kind of birdfeeder, where they have an excuse to be in the presence of someone they admire. You don't have that in a room with just people. You can't sidle up to someone and say the decaf is surprisingly good, don't you think? and start up a conversation. So, they don't get those social experiences. Even in the rooms dedicated to being social, it felt weird and hard to barge into a circle of people just standing around and becoming part of it.

Wearing Blinders and Earmuffs

The disconnect in the virtual reality world wasn't just a cognitive one, it was also perceptual. In the chapter on recognition,

I discussed how much we move to take in information about items in our environment. In the virtual world, the only cues available were the ones designed into the system. Attendees could move, but it was as if they were wearing blinders. Instead of the rods in their eyes responding to the slightest movement in the periphery, their field of view was limited by the screen size and resolution. Levy mentioned multiple times a group gathered to talk, but no one could see or notice others trying to enter the discussion circle. Shoving one's way into a group discussion is rude in real life, but in real life people hear a newcomer coming up behind them or see them out of the corner of their eyes. Then they move to accommodate. At the virtual conference, barging in or creating wild movement to attract attention might be the only hope of joining a conversation – especially when you can't ask someone to pass the milk and sugar.

Phantomatic Machines

> The laws of physiological optics, gravity, and so on, must be faithfully reproduced (unless the subject of the selected vision contravenes this: somebody will want to fly by just "flapping his arms," i.e., against gravitation). Yet apart from the strict deterministic chains of causes and effects mentioned earlier, the vision must also contain a mobile group of processors with a certain degree of freedom: this simply means that the characters appearing in it, our protagonist's phantomatic partners, need to exhibit human characteristics – a (relative) autonomy of speech and action from the actions and words of the protagonist.
>
> – Stanislaw Lem, 1964, "The Phantomatic Machine," in *Summa Technologiae*

The science fiction writer Stanislaw Lem foresaw virtual reality with surprising clarity in the 1960s. His writings in *Summa Technologiae*, finally translated into English in 2013 (seven years after his death), provided a detailed overview of how virtual reality might function, what would need to be represented faithfully, and of course how too much of a good thing could ruin humanity. Lem, like all great writers, was clued into human psychology and knew the importance of engaging the senses when presenting a facsimile of reality.

Video games have always been virtual worlds, but with the advent of VR they can consume the whole of a players' visual, auditory, and tactile experience. I imagine that video games are a touchstone for many readers, as Pew reports that 43 percent of the US population plays video games at least sometimes.[5] It is also a multi-billion dollar business and growing bigger each year.[6]

Every video game has a mental model. That model may or may not match the "real world," and that's fine. We can't fly in our real lives, but it takes little effort to adapt to flying in a virtual world. Some features are so common in games that they are expected: players can move with physics-defying action, such as jumping wild distances, flying, or running up stairs as fast as they can run down them. (My personal attempts in the physical world deem this impossible.) Even time is flexible in a game, with days passing in minutes or minutes passing in hours. Like reading a good book, games are an escape, and a good designer knows that too much gritty realism (such as getting winded after a sprint) would be frustrating verisimilitude. Imagine playing a game where, after 20 minutes of running exploration, the player's avatar just sat down and refused to move. The only example I know of to the contrary was a game designed by the magician Penn Jillette, of Penn and Teller fame. He created the game *Desert Bus* where the player drives a bus from Tucson, AZ to Las Vegas, NV at a top speed of 45 mph.[7] The bus pulls slightly to the right and needs constant correction (and constant, fatiguing attention). The straight road has no decorations. There are no other vehicles. It takes eight hours. The game cannot be paused. And yet, people play it.

It's not important that the mental model match the real world in many ways, but it is important that the game conveys the right mental model for the world of the game and that it stays consistent. Those who signed up for eight hours of *Desert Bus* knew what they were getting into. But lack of consistency was what got the conference attendees outed with their whispering – in some rooms sound *was* attenuated, so only those nearby could hear, but in others a whisper carried over football fields of space. One of the best human attributes is being able to learn and adapt to almost anything, but it has to be consistent.

I spoke with human factors psychologists who worked for years to make video games challenging, playable, and most of all fun: Tim Nichols and Travis Bowles from Microsoft Studios and Celia Hodent from Ubisoft, LucasArts, and then Epic Games. If this sounds like a dream job, you're not alone. And it does involve playing a lot of games. But they needed more than a love of games to be hired – they had to have the scientific background and knowledge to even know *how* to go about making games more fun. Tim and Travis were in the same human factors graduate psychology lab at Georgia Tech that I was in, where we got in trouble for linking the lab computers to play multiplayer games before online games were common. Celia has a PhD in developmental psychology from the University of Paris 5 Sorbonne in France and we met when she worked at Epic Games in North Carolina and have kept in touch as she became the voice of psychology and design in games.[8] Research psychologists like Tim, Travis, and Celia are in demand at game companies because they bring to the job a background in the mental processes, those all-too-human limitations, that were covered in this book, along with the rigorous methods designing studies to know the users (players) and know the task (games).

As Tim said, of his preparation for the job at the American Psychological Association (APA),

> In grad school, I split my time between thinking about the psychology of attention and implicit learning [learning without even knowing you are learning] . . . and more applied problems like telemedicine system design and the design of warnings for older users. Although I knew I didn't want to stay in academia, I soon realized that, for me, there were two awesome things about academic research. One, I loved turning common language questions into testable research hypotheses, and two, I loved poring over data to see whether those hypotheses held up.[9]

The vast worlds of games offered ample opportunities to test hypotheses and then use the results to change the world of the game.

They had some amusing anecdotes about surprises that came about when a game ran counter to a player's mental model. One favorite was a player who steadfastly adhered to all the traffic rules in a driving game designed for fast action, crashes, and the joy of disobeying lights and signs. Nothing about the game changed their mental model of driving. Stop politely for pedestrians! No right on red!

Travis mentioned they were surprised when some well-known games were being redesigned for virtual reality. *Minecraft*, one of the most popular games of the twenty-first century, is a creative building game in a 3-D world only limited by the skill and imagination of the player. It's a "sandbox game" because the game supplies the tools and some rules, but the players create the content (and their own rules, when desired). This can be anything from building a shelter for nighttime protection from monsters to having all the time in the world to try and replicate the Colossus of Rome. It's a little like LEGO in its flexibility with one standard action: players mine for the building resources – after all, it's called *Minecraft*. Typically, a *Minecraft* player sits in front of a screen with a keyboard and a mouse with a first-person view of the world around them. There are well over 100 million players. The creator of this game, is of course, a billionaire.

After *Minecraft* was bought by Microsoft in 2014, there was a move to put the game in the *very* 3-D world of virtual reality. This might not seem like a daunting task as the game was already in three dimensions on the screen. It wasn't like when Super Mario moved from the 2-D scrolling screen to a full 3-D world, with entirely new content and movements. But there were a lot of unknowns to answer, like how big the player should be in relation to the objects in the environment or how should the player move about, since walking with a VR headset was dangerous. The final product was lauded, with *Vice Media* saying, "Minecraft in VR does a few things every developer should copy," so clearly the design and testing worked.[10]

Travis shared a story he told at the MIT Northwest Forum on *The Future of Fun* in 2016 that demonstrated how important understanding mental models was for the development of *Minecraft VR* and how forcefully all our years of experience (and even evolution) affect what we do in a virtual world.

> When we dropped people into *Minecraft* in VR they were fine as long as they were crafting, moving laterally, or climbing up. The weirdness came when the only way to get off a structure was to jump. You can look down, just like you were really standing on a building, and see how far away the ground was. Players hesitated before stepping off, and slowed their approach to edges, even though these short drops wouldn't hurt them in the game at all. The drop looked real enough, even in the cartoonish world of *Minecraft*, to give them pause.

Keep in mind that the player was standing in the middle of their living room, and that "stepping" didn't even require them to move their legs, just use a game controller.

Should falling in VR be designed out of the system? Perhaps not. Fear can also be a feature. Researchers at Clemson University interviewed new *Minecraft VR* players and found that walking off a cliff was something many looked forward to, much like the safe thrill of a roller coaster. One of their participants even built the tallest structure they could, just to jump off of it.[11] Travis confirmed this decision, saying, "A lot of what made *Minecraft VR* an 'experience' was moments like this. Finding new excitement in things that had become routine after many hours of non-VR play. Going underwater, jumping off a ledge, or facing a spider suddenly felt very new, and people reacted." The VR world may differ from the physical one, but consistent rules can be learned, and violating a long-held mental model might offer delight.

Celia and her team also saw playtesters try to use their mental models of the physical world in a virtual one. In the VR game *Bullet Train* by Epic Games, players needed to grab hand-sized rockets out of the air and lob them back at their attacker. Grabbing them was easy in VR, because the player just had to swipe at the incoming rocket and it would be in their hand. The game designers wanted players to immediately throw the rocket back, like a baseball. What they didn't count on was the shape of the rocket – sharp at one end and flat on the other. "Because the pointed end of the rocket was aimed at the player, players tried to turn the rocket around to throw it back," she said, and mimed twisting her arm about. "They ended up in these twisted, awkward and uncomfortable postures, trying to keep hold of the rocket, but point it away from themselves."[12]

Fortunately, capturing undesired behavior during playtesting meant the designers could nudge players in the right direction. They did this by first slowing down the initial rocket attack, so the players had enough time and mental capacity to learn how to catch and throw the rockets. Then, they included a swift but obvious animation. As soon as the rocket was grabbed, it pivoted in the player's hand, so that it was already facing the right direction. Further, the rockets had fire coming out of their back end, making the back of the rocket very different from the front. It was brighter, the fire had motion, and it was a red color not present in the rest of the game world. That conspicuity meant that players could use all their recognition abilities quickly and unconsciously – there was no question about whether the rocket was facing the right way.

Conclusion

Life within technology is still life – we bring with us our human desires, talents, and foibles. But life within technology is a designed life – it's a built world. We build it. We can build upon it, tear it down, rebuild it better. We can make it fit us rather than twisting our (very adaptable) selves to fit it.

We know how to do this by using human factors methods to create human-centered designs. We get to know the user by understanding psychology (attention, motivation, predictable biases, decision-making, creativity) and by understanding culture (experiences, expectations, context). We get to know the tasks by breaking them down, by decomposing them into the smallest of decisions and actions through task analysis. We can analyze the steps, look for redundancies or pressures within the task that prey on our weaknesses, and change them (or offload them onto a computer). We can even categorize the steps of interacting with the world into the types of attention they demand from us and try to prevent them from overwhelming any one of our senses. When it comes to social interactions via technology, whether that be games, videoconferences, or virtual worlds, the introduction to *The Outer Limits* holds true: We can control the horizontal. We can control the vertical. And with principles, input, and testing of the designs, we can make all of the interactions safer, less frustrating, and more enjoyable.

CONCLUSION

Think and do.

– Motto, North Carolina State University, Raleigh, North Carolina

I began this book with a quote from the philosopher Bertrand Russell: "To understand the actual world as it is, not as we should wish it to be, is the beginning of wisdom." I hope by this point you agree with this addendum: it is also the beginning of action and change.

I proposed that there were three steps needed to improve our technological world. The first was to accept that to err is human. To forget is human. To misperceive is human. To fail to notice things right in front of us is human. The World War II colonel demanding, "There will be no more bad landings," can expect to have as much effect as if he decreed that fire should stop being hot. All the proclamations in the world, even the threat of death, can't elevate us beyond our humanity. Human factors professionals sum this sentiment up in one saying: Don't blame the user. This is humans as they are.

The second step was to honestly assess ourselves, through scientific studies, to know what we can and can't expect from human beings. Often, this means we must confront the fact that our capabilities are not all we wish they were. We are irrational, afflicted by influences of which we aren't even aware, making decisions or taking overly optimistic actions that can end in disaster. But all of these attributes have been studied. We have measures of our capabilities and limits, and more knowledge is being discovered all the time. The limits of attention, the ingredients of creativity, well-known biases in our judgements and decisions, how we change our behavior due to payoffs and punishment: all of these form the core of our humanity.

Much of the scientific measurement comes from psychology, especially cognitive psychology and social psychology. Human factors takes that information, the knowledge about people, and applies it to the design and improvement of real systems in the real world. With this knowledge we can predict what people can do, what they can't, and even what they will decide to try and do (before they fail at it).

The last step was to use the knowledge of human capabilities and limitations to change the world around us. Create a world where you don't have to remember that your toddler is in the back seat because, when the circumstances line up, all people can fail to remember. Change the situation. Change the design. You won't be changing the people, at least not without a few more hundred thousand years of evolution. With this knowledge we can design automation to help us perform beyond our meagre human abilities, from flying a spacecraft to even just safely following GPS directions while driving.

What Can We Do?

The university where I work values scientist-practitioners. These are researchers who contribute to understanding basic knowledge about the world, from physics to psychology, and then apply it to immediately improve people's lives. This is codified in the motto of the university, "Think and do." Understand, then act. This book empowers you, the reader, to act.

As a consumer, you can speak with your wallet, demanding that products have a human-centered design and returning them when they don't. You can ask companies to be transparent about their design and testing process. If they know their customers care, they will be more willing to invest in the best practices. You can choose products that are safer because you understand whether they take into account that people *will* make mistakes and design guards to prevent those mistakes from turning into tragedy. You can support companies that produce athletic clothing with reflectors in the places most likely to signal biomotion, and share the importance of reflectivity in the right spots with your friends and family.

In your government, from local to national, you can demand politicians make decisions based on evidence. You can watch for, and call out, incidents of confirmation bias, overreaching decisions based on "fad accidents," and demand transparent and regulated systems for how your country treats people in prisons, detainees, or immigrants seeking asylum. You can insist that the police in your area have training in how to interview children without unintentionally altering or adding to the child's memories. You can have input on the writing of ballot measures to make sure they aren't framed to be misleading. You can have input on laws regarding distracted driving.

Teach your friends and family the importance of good design and good systems, and that they can't depend on just 'being careful,' because distractions can grab their attention when they need it the most. You have an expanse of knowledge now about where humans excel and where they are likely to fail. You can explain the whys to others, keeping them from texting while driving, from thinking a serious memory lapse would never happen to them, and from blaming the victim when another person fails due to a situation asking more from them than the human mind could give. You can watch for biases, like the availability heuristic and confirmation bias, as they sneak their way into the thinking of those around you. Armed with knowledge of confirmation bias, you can purposefully do internet searches for the opposite of what you believe to be true. At the least, you can have more empathy for those who make mistakes, looking to the situation and the other factors that influenced them. Stop blaming victims.

Processes are inherent to all work, especially when more than one person must work together, and now you know how to examine work processes for issues and propose solutions. Use the Gilbreths' time-motion studies as inspiration for how to improve processes at your workplace, whether it be to make a payment system more efficient, speed up reimbursements, or choose the most humane software for meetings. Reduce the friction with technology by changing it or switching to a more usable system. Do employees at your workplace constantly violate workplace rules? You can apply Reason's Swiss cheese model to explain their behavior and have the best chance at changing it. Instead of firing one 'bad apple' after another, you can help them all be good apples. The Gilbreths' dream of 'increased production in shorter time with easier work' can be a reality.

Human factors won't cure cancer. But improving systems and technology for cancer researchers just might. It may feel like we have to deal with the world as it is, but that's not true. We only have to deal with the world as *we* are. Armed with knowledge about the capacity and hazards within the human mind and the methods of design and testing from human factors psychology, we can stop trying in vain to change ourselves or those around us. Instead, we can change our world.

NOTES

Introduction

1 F. B. Gilbreth and E. Carey, *Cheaper by the Dozen* (New York: Thomas Y. Crowell Co., 1948).
2 Ibid.
3 The anecdote was provided by: A. Chapanis, *The Chapanis chronicles: 50 years of human factors research, education and design* (Santa Barbara: Aegean, 1999). A clearly written overview may be found in S. A. Ruffin, "Human factors research: Meshing pilots with planes," in R. P. Hallion (ed.), *NASA's contributions to aeronautics, Vol. 2* (Washington, DC: NASA, 2010).
4 Fitts and Jones wrote two reports from the same data: (1) P. M. Fitts and R. E. Jones, *Analysis of factors contributing to 460 "pilot error" experiences in operating aircraft controls* (Report No. TSEAA-694-12; Dayton, OH: Aero Medical Laboratory, US Air Force, 1947) and (2) *Psychological aspects of instrument display. Analysis of 270 "pilot-error" experiences in reading and interpreting aircraft instruments* (Report No. TSEAA-694-12A; Dayton, OH: Aero Medical Laboratory, US Air Force, 1947).
5 This anecdote was included in Chapanis, *The Chapanis chronicles*.
6 R. H. Cameron, *Training to fly: Military flight training, 1907–1945* (Washington, DC: Air Force History and Museums Program, 1999).

Chapter 1

1 All dialogue and timing information from Sullenberger, Skiles, and Harten comes from the transcription of cockpit voice recorder from US Airways Flight 1549, January 15, 2009. Full transcript can be found at www.ntsb.gov/investigations/AccidentReports/Reports/AAR1003.pdf and further transcripts, including discussion at each communication tower, can be found at the US FAA repository (www.faa.gov/data_research/accident_incident/2009-01-15/, retrieved March 28, 2021).
2 Information on Captain Sullenberger's biography comes from his own website: www.sullysullenberger.com/about, and information on co-pilot Jeffery Skiles biography comes from his testimony to the US House of Representatives Congressional Subcommittee on Aviation, February 24, 2009. The transcript of this congressional

hearing is available at www.govinfo.gov/content/pkg/CHRG-111hhrg47866/pdf/CHRG-111hhrg47866.pdf (retrieved March 28, 2021).

3 S. Johnson, *Future perfect: The case for progress in a networked age* (London: Penguin, 2013).

4 Associated Press, *NTSB: Sully could have made it back to LaGuardia*, CBS Interactive, Inc. (May 4, 2010). Retrieved March 24, 2021 from www.cbsnews.com/news/ntsb-sully-could-have-made-it-back-to-laguardia/.

5 National Transportation Safety Board, *Loss of thrust in both engines after encountering a flock of birds and subsequent ditching on the Hudson River* (Publication NTSB/AAR-10/03, PB2010-910403; Washington, DC: NTSB, 2010), www.ntsb.gov/investigations/AccidentReports/Reports/AAR1003.pdf, retrieved March 24, 2021.

6 ICAO, "Celebrating TAWS 'saves': But lessons still to be learnt," *International Civil Aviation Organization* (n.d.), www.icao.int/safety/fsix/Library/TAWS%20Saves%20plus%20add.pdf, retrieved March 24, 2021.

7 National Transportation Safety Committee, *Aircraft accident investigation report: Sukhoi civil aircraft company, Sukhoi RRJ-95B; 97004* (Publication KNKT.12.05.04; Republic of Indonesia, 2012). Retrieved March 24, 2021 from http://web.archive.org/web/20130123112752/http://www.dephub.go.id/knkt/ntsc_aviation/baru/Final%20Report_97004_Release.pdf.

8 Committee for Investigation of National Aviation Accidents, *Final report from the examination of the aviation accident no. 192/2010/11 involving the Tu-154M airplane, tail number 101, which occurred on April 10th, 2010 in the area of the SMOLENSK NORTH airfield* (2011). Retrieved March 24, 2021 from http://wayback.archive-it.org/all/20120906032711/http://mswia.datacenter-poland.pl/FinalReportTu-154M.pdf.

9 National Transportation Safety Board – Office of Administrative Law Judges, *Public hearing in the matter of the landing of US Airways flight 1549, N106US, in the Hudson River, Weehawken, New Jersey, January 15th, 2009* (Free State Reporting, Inc., June 9, 2009). Retrieved March 24, 2021 from www.exosphere3d.com/pubwww/pdf/flight_1549/ntsb_docket/422295.pdf.

10 L. Shiner, "Sully's tale," *Air and Space Magazine* (Smithsonian Institution, Washington, DC, 2009). Retrieved March 24, 2021 from www.airspacemag.com/as-interview/aamps-interview-sullys-tale-53584029/.

11 Transcribed from video of *The Daily Show with Jon Stewart*, season 14, episode 132, originally aired October 13, 2009 on Comedy Central.

12 The following are good starting papers for those interested in knowing more about theory of mind: (1) I. A. Apperly, "What is 'theory of mind'? Concepts, cognitive processes and individual differences," *Quarterly Journal of Experimental Psychology*, 65 (2012), 825–839; (2) J. W. Astington and A. Gopnik, "Theoretical explanations of children's understanding of the mind," *British Journal of Developmental Psychology*, 9 (1991), 7–31; and (3) D. W. Griffin W. and L. Ross, "Subjective construal, social inference, and human misunderstanding," in M. Zanna (ed.), *Advances in experimental social psychology*, vol. 24 (New York: Academic Press, 1991), pp. 319–359.

13 E. L. Newton, *The rocky road from actions to intentions* (unpublished doctoral dissertation, Stanford University, 1990). Also frequently referenced as *Overconfidence in the communication of intent: Heard and unheard melodies.*

14 A. C. McLaughlin, J. Ward, and B. W. Keene, "Development of a veterinary surgical checklist," *Ergonomics in Design*, 24(4)(2016), 27–34.

15 A. Gawande, *The checklist manifesto* (New York: Picadur, 2010).

Chapter 2

1 Archive.org maintains the full text of all emails sent to and from Governor Rick Snyder during this time. The emails may be accessed through http://archive.org/stream/snyder_flint_emails/Staff_10_djvu.txt (retrieved April 20, 2021).
2 Ibid.
3 T. M. Olson et al., "Forensic estimates of lead release from lead service lines during the water crisis in Flint, Michigan," *Environmental Science & Technology Letters*, 4 (9)(2017), 356–361.
4 The Flint Water Advisory Task Force, appointed by Governor Rick Snyder, issued a final report in 2016 containing this information that the MDEQ had misread the rule. The final report can be found online at: www.michigan.gov/documents/snyder/FWATF_FINAL_REPORT_21March2016_517805_7.pdf (retrieved April 20, 2021).
5 Environmental Protection Agency (EPA), *Management weaknesses delayed response to Flint water crisis* (Report No. 18-P-0221, Office of Inspector General, US Environmental Protection Agency, 2018).
6 A full version of the final report by Miguel del Toral, finalized in 2015, can be found in the archives of the EPA. This fascinating report contains his analysis of the situation in Flint complete with photos of the pipes and surrounding areas where high lead levels were detected. It can be read at www.epa.gov/sites/production/files/2015-11/documents/transmittal_of_final_redacted_report_to_mdeq.pdf (retrieved April 20, 2021).
7 The complete and personal story of Mona Hanna-Attisha's attempts to illuminate and solve the lead poisoning of Flint can be found in her 2018 book, *What the eyes don't see* (New York: One World).
8 These facts and timeline were gathered from Donovan Hohn's *New York Times* article, "Flint's water crisis and the 'troublemaker' scientist," published on August 16, 2016.
9 This quote comes from an email from Geralyn Lasher at the MDHHS to Dennis Muchmore, Chief of Staff for Governor Rick Snyder, on September 25, 2015. It is archived in the 2018 book *The great water: A documentary history of Michigan*, ed. Matthew R. Thick (East Lansing: Michigan State University Press).
10 Numerous emails from the archive linked in the first note for this chapter refer to the poor aesthetics of the water but declare it is safe. In particular, a *Water quality report* by the company Veolia was sent on March 12, 2015 and stated that "the public has also expressed its frustration over discolored and hard water. Those aesthetic issues have understandably increased the level of concern about the safety of the water. The review of the water quality records during the time of Veolia's study shows the water to be in compliance with State and Federal regulations, and, based on those standards, the water is considered to meet drinking water requirements." This report can be found in the email archive accessed through http://archive.org/stream/snyder_flint_emails/Staff_10_djvu.txt (retrieved April 20, 2021).
11 A. Sorkin, "The contempt that poisoned Flint's water," *The New Yorker* (January 22, 2016).
12 Ibid.
13 Email from Dennis Muchmore, Chief of Staff to Governor Rick Snyder. This email is archived in *The great water*.
14 This quote came from an email to reporter Ron Fonger from Brad Wurfel (MDEQ) on September 9, 2015. The quote was repeated in multiple news sources and can be found archived in Katrinell M. Davis's book, *Tainted tap: Flint's journey from crisis to recovery* (Durham, NC: University of North Carolina Press, 2021).

15 See A. Tversky and D. Kahneman, "The framing of decisions and the psychology of choice," *Science*, 211(4481)(1981), 453–458.
16 This amendment to the Missouri constitution was proposed by the 97th General Assembly CCS No. 2 SS HCS HJR Nos. 11 and 7. It appeared on the August 5, 2014 primary election ballot. You can learn more about whether such confusing language is intentional on ballots from this *Washington Post* article: J. Fuller, "Why are ballot measures so darn confusing? Because they are supposed to be," *The Washington Post* (2014), www.washingtonpost.com/news/the-fix/wp/2014/08/05/why-are-ballot-measures-so-darn-confusing-because-they-are-supposed-to-be/ (retrieved March 29, 2018).
17 I. P. Levin and G. J. Gaeth, "How consumers are affected by the framing of attribute information before and after consuming the product," *Journal of Consumer Research*, 15(3)(1988), 374–378.
18 C. A. Bigman, J. N. Cappella, and R. C. Hornik, "Effective or ineffective: Attribute framing and the human papillomavirus (HPV) vaccine," *Patient Education and Counseling*, 81 (2010), 70–76.
19 V. F. Reyna, C. F. Chick, J. C. Corbin, and A. N. Hsia, "Developmental reversals in risky decision making: Intelligence agents show larger decision biases than college students," *Psychological Science*, 25(1)(2014), 76–84.
20 S. Kim, D. Goldstein, L. Hasher, and R. T. Zacks, "Framing effects in younger and older adults," *The Journals of Gerontology. Series B, Psychological Sciences and Social Sciences*, 60(4)(2005), 215–218.
21 US Secretary of Defense Donald Rumsfeld delivered this line in a February 12, 2002 Department of Defense news briefing about the lack of evidence for Weapons of Mass Destruction in Iraq. The quote in context is, "Reports that say that something hasn't happened are always interesting to me, because as we know, there are known knowns; there are things we know we know. We also know there are known unknowns; that is to say we know there are some things we do not know. But there are also unknown unknowns – the ones we don't know we don't know. And if one looks throughout the history of our country and other free countries, it is the latter category that tends to be the difficult ones."
22 The Director of the US Central Intelligence Agency (CIA), George J. Tenet, distributed a statement and press release on August 11, 2003. This press release goes into detail about the history of US intelligence on Iraq and concludes that he stands behind the assessment that Iraq was developing WMD. This press release is archived at http://nsarchive2.gwu.edu/NSAEBB/NSAEBB80/Press%20Release.htm. A statement a month later by UN Chief Weapons Inspector David Kay to the US House Committee on Intelligence offered similar content and specifically mentioned, "We have not yet found stocks of weapons, but we are not yet at the point where we can say definitively either that such weapon stocks do not exist or that they existed before the war and our only task is to find where they have gone." This statement is archived at http://nsarchive2.gwu.edu/NSAEBB/NSAEBB80/Statement%20on%20the%20Interim%20Progress%20Report%20on%20the%20Activities%20of%20the%20Iraq%20Survey%20Group.htm. By 2004, Kay publicly acknowledged there were no WMD to be found in Iraq and encouraged UK and US leaders to acknowledge the same. Kay was the rare human who incorporated new evidence, even in an emotional situation, and reversed his thoughts and opinions.
23 J. D. Stuster, "The Iraq syndrome," *Foreign Policy Magazine* (March 19, 2013).
24 C. K. Morewedge et al., "Debiasing decisions: Improved decision making with a single training intervention," *Policy Insights from the Behavioral and Brain Sciences*, 2(1)(2015), 129–140.

25 R. J. Heuer, *How does analysis of competing hypotheses (ACH) improve intelligence analysis?* (Reston, VA: Pherson Associates, LLC, 2005).

Chapter 3

1 V. P. Glăveanu and M. Taillard, "Difficult differences pave the creative road from diversity to performance," *European Management Journal*, 36(6)(2018), 671–676.
2 A. S. McKay, M. Karwowski, and J. C. Kaufman, "Measuring the muses: Validating the Kaufman domains of creativity scale (K-DOCS)," *Psychology of Aesthetics, Creativity, and the Arts*, 11(2)(2017), 216.
3 The original paper on this comes from the mid-twentieth century: A. S. Luchins, "Mechanization in problem solving: The effect of Einstellung," *Psychological Monographs*, 54(6)(1942), 1–95. For a well-written and entertaining modern summary, see: M. Bilalić and P. McLeod, "Why good thoughts block better ones," *Scientific American*, 310(3)(2014), 74–79.
4 To check my memory of this tour, I called up Shelby Mattice, the Curator of the Bronck Museum in Greene County, NY. She was the tour guide of my tour a decade ago and confirmed the leaky design of the roof. However, she also provided additional historical perspective. Being Dutch in early Colonial America meant being in a cultural minority, where preservation of their history and ways of living was more important than pursuing better roof designs. She pointed out that the English methods of roofing were also leaky at the time, so a better method was not a clear choice.
5 Many sources refer to Apollo 13 as "the successful failure" with no individual attribution. NASA uses the title on their history website at www.nasa.gov/centers/marshall/history/apollo/apollo13/index.html (retrieved April 20, 2021).
6 Details of this story came from the transcripts of the Apollo 13 mission created by W. David Woods, Johannes Kemppanen, Frank O'Brien, Alexander Turhanov, and Lennox J. Waugh. These transcripts are maintained at http://history.nasa.gov/afj/ap13fj/ (retrieved April 20, 2021).
7 All quotes from John Stuart Mill come from J. S. Mill, *The principles of political economy: With some of their applications to social philosophy*, The Collected Works of John Stuart Mill (1848) (ed. J. M. Robson; Toronto: University of Toronto Press, 1965), vol. 3, book V, ch. XVII, sec. 3.
8 The research supporting this is not without controversy. The reference comes from S. Danziger, J. Levav, and L. Avnaim-Pesso, "Extraneous factors in judicial decisions," *Proceedings of the National Academy of Sciences*, 108(17)(2011), 6889–6892. Though 1,000 judicial decisions were analyzed, it was only across eight judges. The authors addressed other criticisms of their analysis in this response: Danziger, Levav, and Avnaim-Pesso, "Reply to Weinshall-Margel and Shapard: Extraneous factors in judicial decisions persist," *Proceedings of the National Academy of Sciences*, 108(42)(2011), e834–e834.
9 H. T. Neprash and M. L. Barnett, "Association of primary care clinic appointment time with opioid prescribing," *JAMA Network Open*, 2(8)(2019), e1910373–e1910373.
10 C. A. Anderson et al., "Violent video game effects on aggression, empathy, and prosocial behavior in eastern and western countries: A meta-analytic review," *Psychological Bulletin*, 136(2)(2010), 151–173.
11 A. R. Damasio, "The somatic marker hypothesis and the possible functions of the prefrontal cortex," *Philosophical Transactions of the Royal Society of London. Series B: Biological Sciences*, 351(1346)(1996), 1413–1420.

12 S. Pershall, *Loud in the house of myself: Memoir of a strange girl* (New York: W.W. Norton & Company, 2012).
13 J. Schroeder et al., "Pocket skills: A conversational mobile web app to support dialectical behavioral therapy," in *Proceedings of the 2018 CHI Conference on Human Factors in Computing Systems* (ACM, 2018), pp. 1–15.
14 S. F. Austin et al., "Mobile app integration into dialectical behavior therapy for persons with borderline personality disorder: Qualitative and quantitative study," *JMIR Mental Health*, 7(6)(2020), e14913.
15 A. R. Damasio, *Descartes' error* (New York: Random House, 2006).
16 M. L. Resnick, "The effect of affect: Decision making in the emotional context of health care," in *2012 Symposium on Human Factors and Ergonomics in Health Care* (New York: Sage Publishing, 2012), pp. 39–44.
17 All alarm statistics came from this report: Association for the Advancement of Medical Instrumentation (AAMI), *AAMI clinical alarms: 2011 summit*, retrieved April 26, 2019 from www.aami.org/publications/summits/2011_Alarms_Summit_publication.pdf.
18 M. Cvach, "Monitor alarm fatigue: An integrative review," *Biomedical Instrumentation & Technology*, 46(4)(2012), 268–277. These two papers are a good introduction to this literature: (1) Glăveanu and Taillard, "Difficult differences pave the creative road," and (2) J. Wang et al., "Team creativity/innovation in culturally diverse teams: A meta-analysis," *Journal of Organizational Behavior*, 40 (6)(2019), 693–708.
19 E. Mannix and M. A. Neale, "What differences make a difference? The promise and reality of diverse teams in organizations," *Psychological Science in the Public Interest*, 6(2)(2005), 31–55.
20 K. W. Phillips, "How diversity makes us smarter," *Scientific American* (October, 2014).

Chapter 4

1 K. Hwang, "'Because of a stupid cellphone': Regret, anger and pain after distracted driving crashes," *Indianapolis Star* (January 21, 2020). Quotes come from a video interview with the *Indianapolis Star*, which can be watched at www.indystar.com/story/news/local/transportation/2020/01/31/distracted-driving-crash-victims-share-stories-hardship-loss/4587525002/ (retrieved April 20, 2021).
2 G. A. Fitch et al., *The impact of hand-held and hands-free cell phone use on driving performance and safety-critical event risk (Report No. DOT HS 811 757)* (Washington, DC: National Highway Traffic Safety Administration, April 2013).
3 D. L. Strayer and W. A. Johnston, "Driven to distraction: Dual-task studies of simulated driving and conversing on a cellular phone," *Psychological Science*, 12 (2001), 462–466.
4 Ibid. (experiment 2).
5 F. A. Drews, M. Pasupathi, and D. L. Strayer, "Passenger and cell phone conversations in simulated driving," *Journal of Experimental Psychology: Applied*, 14(4) (2008), 392–400.
6 Fitch et al., *The impact of hand-held and hands-free cell phone use*.
7 The original article is: A. D. Baddeley and G. Hitch, "Working memory," in *Psychology of Learning and Motivation* (New York: Academic Press, 1974), vol. 8, pp. 47–89). Baddeley continued work on this model for the next few decades with an update to include an *episodic buffer*: A. Baddeley, "The episodic buffer:

A new component of working memory?," *Trends in Cognitive Sciences*, 4(11)(2000), 417–423.

8 C. D. Wickens, "Multiple resources and performance prediction," *Theoretical Issues in Ergonomic Science*, 3(2)(2002), 159–177.

9 C. D. Wickens, "Multiple resources and mental workload," *Human Factors*, 50(3)(2008), 449–455.

10 J. Abumrad and R. Krulwich, *Gut feelings* (Radiolab Podcast, WNYC Studios, New York, 2012).

11 The 2002 paper mentioned earlier, Wickens, "Multiple resources and performance prediction," contains an explanation of how to calculate task conflict levels.

12 D. G. Hoecker et al., "Man-machine design and analysis system (MIDAS) applied to a computer-based procedure-aiding system," in *Proceedings of the Human Factors and Ergonomics Society Annual Meeting*, vol. 38, no. 4 (Los Angeles: SAGE Publications, 1994), pp. 195–199).

13 Information and updates about MIDAS are maintained on NASA's website: http://humansystems.arc.nasa.gov/groups/midas/design/new_midas_5.html (retrieved April 20, 2021).

Chapter 5

1 All quotes from Lynn Hill come from her answers to the Reddit.com AMA ("Ask Me Anything") available at: www.reddit.com/r/IAmA/comments/33n893/i_am_lynn_hill_the_first_female_climber_to_free/ (retrieved May 20, 2021).

2 D. Kahneman, *Thinking, fast and slow* (New York: Farrar, Straus and Giroux, 2013).

3 A reference and a fun fact about your author. The term "controlled processing" came into use via this paper: W. Schneider and R. M. Shiffrin, "Controlled and automatic human information processing: I. Detection, search, and attention," *Psychological Review*, 84(1)(1977), 1. The authors are my great and great-great academic grandfathers (my PhD advisor's advisor's advisor and that advisor's advisor).

4 The name for this kind of test is a Stroop Test, named after J. Ridley Stroop and his 1935 dissertation work where he had people try to name the color of ink that color names were printed in (e.g., BLUE was printed in yellow ink, and it took longer and was more difficult to say "yellow" than one might think). Stroop tests now have a long history of being used to detect issues with attention, including comparing a person's baseline performance to when they are in an extreme environment (such as the top of Mount Everest).

5 G. Wolf, "Steve Jobs: The next insanely great thing," *WIRED* (February 1, 1996).

6 S. B. Kaufman, "The real link between creativity and mental illness," *Scientific American*, http://blogs.scientificamerican.com/beautiful-minds/the-real-link-between-creativity-and-mental-illness/ (retrieved March 31, 2020).

7 S. B. Kaufman, "The creative gifts of ADHD," *Scientific American*, http://blogs.scientificamerican.com/beautiful-minds/the-creative-gifts-of-adhd/ (retrieved March 31, 2020).

8 The details of the story came from the arrest warrant for Kyle Seitz, made public on November 12, 2014. The original warrant is listed in the State of Connecticut Superior court, JD-CR-71 Rev. 3-11, Ridgefield Police Department, Police Case number, 140001 0124.

9 K. Wallace, *A grieving mother's mission to stop hot car deaths*, CNN (September 2, 2014), www.cnn.com/2014/07/31/living/hot-car-deaths-mothers-mission/index.html (retrieved May 20, 2021).

10 S. Hostin, *I, too, left my child in a hot car*, CNN (June 26, 2014), www.cnn.com/2014/06/25/opinion/hostin-hot-car-child (retrieved May 20, 2021).

11 A. C. Vollers, *Rubber band device invented by 11-year-old could keep parents from leaving children in hot cars*, AL.com (June 27, 2014), www.al.com/living/2014/06/rubber_band_device_invented_by.html (retrieved May 20, 2021).

12 C. D. Wickens and C. M. Carswell, "The proximity compatibility principle: Its psychological foundation and relevance to display design," *Human Factors*, 37(3) (1995), 473–494.

13 Injuries reported in: (1) K. C. Chung and M. J. Shauver, "Table saw injuries: epidemiology and a proposal for preventive measures," *Plastic and Reconstructive Surgery*, 132(5)(2013), 777e, and (2) amputations reported by Inc.com from a Consumer Product Safety Commission estimate: M. Newsome (January 24, 2007). He took on the whole power-tool industry: "Why wasn't anyone else interested in building a safer saw?," *Inc. Magazine.*

14 July 26, 2017 Comments of National Consumers League to the US Consumer Product Safety Commission on *Table saw blade contact injuries; Notice of proposed rulemaking; Requests for comments and information*, http://nclnet.org/wp-content/uploads/pdf/Table_Saw_Safety_Comments_-July_26.pdf (retrieved May 20, 2021).

15 This book section required much more research than I expected to evaluate the claims made by the Power Tool Institute, the Consumer Product Safety Commission (CPSC), SawStop, and the National Consumers League. I do not address the claims that mandating safety technology might lead to monopoly or other legal issues because my focus is on the safety of design for humans. The lawyers can worry about the legal issues. The claim that including "flesh sensing" safety measures in a table saw will result in careless and complacent users comes from the PTI website: "Data supplied by SawStop concerning the number of table saw units sold and the number of reported blade contact incidences, demonstrates that operators are nearly five times more likely to contact the saw blade of a SawStop saw as opposed to the operator of a conventional table saw. Logic dictates that this increase in accident rate on SawStop saws is due primarily to a user's decision to use the blade guard less frequently or not at all due to a 'sense of security' in having the SawStop flesh-sensing technology on the saw" (www.powertoolinstitute.com/pti-pages/it-table-saw-facts.asp, retrieved May 21, 2021). I thought the claim that accidents were higher with SawStop was worth investigating, and found this rebuttal from the editors of *Woodshop News*, a trade magazine published since 1986 (October 26, 2011): "The PTI says users of SawStop saws 'are nearly five times more likely to contact the SawStop's saw blade as opposed to an operator of a conventional saw.' That allegation is the result of a math error by the PTI where they assumed that accidents on SawStop saws all occurred in one year, when in fact they occurred over five years. The data actually shows that the accident rate on table saws without injury mitigation technology is approximately 0.7% per year for non-workplace accidents. The most comparable saw equipped with injury mitigation technology is SawStop's contractor saw, and the accident rate on SawStop's contractor saw is also 0.7% per year." This would mean that the accident rate is the same with or without the SawStop, making the most interesting data the *severity* of the accidents, since even a small cut is still considered an accident. I was not able to find those data.

Chapter 6

1 G. Saji, "Safety goals for seismic and tsunami risks: Lessons learned from the Fukushima Daiichi disaster," *Nuclear Engineering and Design*, 280(2014), 449–463.
2 G. Gavett, *Voices from the inside: Fukushima's workers speak*, PBS Frontline (March 11, 2012), www.pbs.org/wgbh/frontline/article/voices-from-the-inside-fukushimas-workers-speak/ (retrieved May 21, 2021). Quotes taken from the PBS Frontline episode *Inside Japan's nuclear meltdown*, which first aired February 28, 2012.
3 J. Reason, "The contribution of latent human failures to the breakdown of complex systems," *Philosophical Transactions of the Royal Society of London. Series B, Biological Sciences*, 327(1241)(1990), 475–484.
4 All details and quotes from Chernobyl workers regarding the state of the USSR and nuclear power come from G. Medvedev and E. Rossiter (trans.), *The truth about Chernobyl* (translated from the Russian ed.) (London: I.B.Tauris, 1991).
5 I highly recommend the following book as a practical and approachable read to the radical idea that human error doesn't exist except as it is created by the systems we thrust people into: S. Dekker, *The field guide to understanding "human error"* (3rd ed.; Hampshire: Ashgate, 2014).
6 Quote from a TASS dispatch and press release from government officials in the USSR that was broadcast on television there.
7 W. F. Schulz, *In our own best interest: How defending human rights benefits us all* (Boston, MA: Beacon Press, 2001).
8 Analysis of the Fukushima accident regarding TEPCO and the Japanese government came from: The Initiative for Global Environmental Leadership (IGEL), *Special report: Disasters, leadership and rebuilding – tough lessons from Japan and the U.S.* (2013), http://d1c25a6gwz7q5e.cloudfront.net/reports/2013–10–01-Disasters-Leadership-Rebuilding.pdf (retrieved October 11, 2020).
9 N. Kulish and N. Clark, "Germanwings crash exposes history of denial on risk of pilot suicide," *New York Times* (April 18, 2015).
10 A. Johnson, *Airlines adopt two-in-the-cockpit rule after Germanwings crash*, NBC News (March 26, 2015), www.nbcnews.com/storyline/german-plane-crash/airlines-worldwide-adopt-two-in-the-cockpit-rule-after-germanwings-n331041 (retrieved May 21, 2021). This rule was later eliminated by many countries, such as in Canada (described in this news story): A. Burke, *Rule requiring airlines to keep 2 crew in cockpit at all times lifted by Transport Canada*, CBC (June 16, 2017), www.cbc.ca/news/canada/ottawa/transport-canada-two-flight-crew-cockpit-1.4164592 (retrieved May 21, 2021).
11 S. A. Shappell and D. A. Wiegmann, *The human factors analysis and classification system – HFACS. DOT/FAA/AM-00/7* (2000).
12 The full report can be found from the Wharton School at the University of Pennsylvania: http://knowledge.wharton.upenn.edu/article/lessons-leadership-fukushima-nuclear-disaster/ (retrieved May 20, 2021).
13 These and all other numerical COVID-19 statistics can be found at the World Health Organization (WHO) website: http://covid19.who.int/ (retrieved May 20, 2021).
14 Information regarding Ohio's response to the early pandemic came from the *Executive orders by the Governor 2020-01D* through *Executive Order 2020-10D*.
15 State comparisons of cases per 100,000 residents came from the US Centers for Disease Control (CDC) databases, accessible at http://covid.cdc.gov/covid-data-tracker (retrieved May 20, 2021).
16 The video of Governor Ivey's statement was carried by many news organizations. It can be viewed online as recorded by AL.com at www.youtube.com/watch?v=

jZ3XOSmWeOo&ab_channel=GovernorKayIveyGovernorKayIveyVerified (retrieved May 20, 2021).

17 R. L. Haffajee and M. M. Mello, "Thinking globally, acting locally – The US response to COVID-19," *New England Journal of Medicine*, 382(22)(2020), e75.

18 D. Thompson, "What's behind South Korea's COVID-19 exceptionalism?," *The Atlantic* (May 6, 2020).

19 COVID-19 Dashboard: South Korea, WHO (2020), http://covid19.who.int/region/wpro/country/kr (retrieved October 12, 2020).

20 The editors, "Dying in a leadership vacuum," *New England Journal of Medicine*, 383(October 8, 2020), 1479–1480.

Chapter 7

1 The original interview aired on NPR in 1986. Interview by Daniel Zwerdling on the radio program *All things considered*. In 2012, soon after his death, the anonymous engineer was revealed to be Roger Boisjoly, who had later brought a lawsuit against the company (Morton Thiokol, Inc.) who approved the launch.

2 The original report, by Leon Ray, is titled *MSFC's SRM Clevis Joint Leakage Study, October 21, 1977*, and can be found as archived in the *Report of the Presidential Commission on the space shuttle Challenger accident*, vol. 5 index, hearings of the Presidential Commission on the space shuttle challenger accident: February 26, 1986 to May 2, 1986.

3 The quotes and information regarding the decisions made after the launch of the Columbia all come from the official *Report of Columbia Accident Investigation Board, Volume I*, released August 26, 2003. It is archived on NASA's website: www.nasa.gov/columbia/home/ (retrieved May 20, 2021).

4 Information on the timeline and events of the Deepwater Horizon oil spill came from: Deepwater Horizon Study Group, *Final report on the investigation of the Macondo well blowout* (Center for Catastrophic Risk Management, 2011), http://ccrm.berkeley.edu/pdfs_papers/bea_pdfs/dhsgfinalreport-march2011-tag.pdf (retrieved March 3, 2020).

5 National Academy of Engineering & National Research Council, *Macondo well Deepwater Horizon blowout: Lessons for improving offshore drilling safety* (Washington, DC: National Academies Press, 2012).

6 Ibid.

7 A. Barnett, "Aviation safety: A whole new world?," *Transportation Science*, 54(1) (2020), 84–96.

8 M. Roser, *Oil spills*, OurWorldInData.org. (2013), ourworldindata.org/oil-spills (retrieved May 24, 2021).

9 P. G. Zimbardo, "On the ethics of intervention in human psychological research: With special reference to the Stanford prison experiment," *Cognition*, 2(1973), 243–256.

10 Phillip Zimbardo did learn this lesson well, and saw the similarities in the example used in this chapter: the Abu Ghraib prison. He covers the topic in his book, *The Lucifer effect: Understanding how good people turn evil* (New York: Random House, 2007).

11 R. Leung, *Abuse of Iraqi POWs by GIs probed: 60 Minutes II has exclusive report on alleged mistreatment*, CBS News (April 27, 2004).

12 R. Leung, *Abuse at Abu Ghraib: Dan Rather has details of one man who died in the custody of Americans*, CBS News (May 5, 2004).

13 Interview with B. G. Janis Karpinski included as an annex to the official report: A. Taguba, *U.S. Army 15-6 Report of abuse of Prisoners in Iraq* (US Department of Defense, May 2004).
14 Taguba, *U.S. Army 15–6 Report of abuse of prisoners in Iraq.*
15 G. W. Bush. *Humane treatment of Taliban and al Qaeda detainees*, White House Memo ((February 7, 2002), The White House, Washington, DC.
16 Committee on Armed Services, *S. PRT. 110-54 inquiry into the treatment of detainees in U.S. custody*. Report of the Committee on Armed Services, United States Senate, 110th Congress, 2nd Session (November 20, 2008).
17 R. Kaplan, *John McCain stands out in defense of CIA torture report release*, *Face the Nation*, CBS News (December 14, 2014).
18 A collection of citations for this work are: (1) S. Shappell and D. Wiegmann, "A methodology for assessing safety programs targeting human error in aviation," *The International Journal of Aviation Psychology*, 19(3)(2009), 252–269; (2) S. A. Shappell and D. A. Wiegmann, *The human factors analysis and classification system – HFACS. DOT/FAA/AM-00/7* (2000); and (3) D. A. Wiegmann and S. A. Shappell, *A human error approach to aviation accident analysis: The human factors analysis and classification system* (New York: Routledge, 2017).
19 United Nations Office on Drugs and Crime (UNODC), "Rules on solitary confinement," in *The United Nations standard minimum rules for the treatment of prisoners* (2015), www.unodc.org/documents/justice-and-prison-reform/Nelson_Mandela_Rules-E-ebook.pdf (retrieved September 8, 2020).
20 J. Abumrad and R. Krulwich, *New nice*, *Radiolab* (October 19, 2015), WNYC Studios.
21 S. B. Paletz et al., "Socializing the human factors analysis and classification system: Incorporating social psychological phenomena into a human factors error classification system," *Human Factors*, 51(4)(2009), 435–445.
22 T. Richissin, "Soldiers' warnings ignored," *Baltimore Sun* (May 9, 2004), www.baltimoresun.com/news/bal-te.guard09may09-story.html (retrieved April 3, 2020).
23 Ibid.
24 PBS, The torture question: Episode 1, Frontline (October 18, 2005).
25 Richissin, "Soldiers' warnings ignored."
26 PBS, The torture question: Episode 1.
27 Interview with B. G. Janis Karpinski.

Chapter 8

1 The original reference for signal detection theory is: D. M. Green and J. A. Swets, *Signal detection theory and psychophysics*, vol. 1 (New York: Wiley, 1966). For a more modern overview, see: C. D. Wickens et al., "Signal detection and absolute judgement," in *Engineering Psychology and Human Performance, Fifth Ed.* (New York: Routledge, 2021). For the most approachable overview, see J. A. Swets, R. M. Dawes, and J. Monahan, "Better decisions through science," *Scientific American*, 283(4)(2000), 82–87.
2 The story of the bear-saboteur was uncovered via the US Freedom of Information Act requests many decades later and can be found in: S. D. Sagan, *The limits of safety* (Princeton, NJ: Princeton University Press, 1993). The last bit to the story was that the wrong alarm sounded at another military base when the bear-saboteur was discovered, one that announced nuclear war had begun.

3 N. Davis, "Soviet submarine officer who averted nuclear war honoured with prize," *The Guardian* (October 27, 2017), www.theguardian.com/science/2017/oct/27/vasili-arkhipov-soviet-submarine-captain-who-averted-nuclear-war-awarded-future-of-life-prize (retrieved May 5, 2021).
4 P. Askenov, *Stanislav Petrov: The man who may have saved the world*, BBC News (September 26, 2013).
5 P. Smith, *The latest TSA embarrassment* (June 15, 2015), http://askthepilot.com/tsa-failure/ (retrieved May 5, 2021).
6 CBS, *Passenger passes TSA, boards plane with loaded gun* (November 16, 2015), www.cbs46.com/story/30523628/passenger-on-plane-with-gun-goes-unnoticed-by-tsa#ixzz3rgSD5nWK (retrieved May 26, 2021).
7 J. M. Wolfe et al., "Prevalence effects in newly trained airport checkpoint screeners: Trained observers miss rare targets, too," *Journal of Vision*, 13(3)(2013), 1–9.
8 Quoted text comes from Rapiscan's website advertising their Threat Image Projection software, www.rapiscansystems.com/en/products/rapiscan-threat-image-projection (retrieved May 26, 2021).
9 M. Anteby and C. K. Chan, "A self-fulfilling cycle of coercive surveillance: Workers' invisibility practices and managerial justification," *Organization Science*, 29(2) (2018), 247–263.
10 B. Peterson, "Inside job: My life as an airport screener," *Conde Nast Traveler* (2007), www.cntraveler.com/stories/2007-02-20/inside-job-my-life-as-an-airport-screener (retrieved April 1, 2019).
11 D. Harris, "How to really improve airport security," *Ergonomics in Design*, 10(1) (2002), 17–22.
12 For an overview of automation concepts and issues, see: R. Parasuraman and V. Riley, "Humans and automation: Use, misuse, disuse, abuse," *Human Factors*, 39(2)(1997), 230–253.
13 Although she has likely told this story numerous times, I heard it when attending her presentation at the Duke Robotics Student Symposium (DSSS), Duke University, Durham, NC, on March 28, 2016.
14 For more scientific coverage of the effects of reward and punishment, see: (1) P. Madhavan et al., "The role of incentive framing on training and transfer of learning in a visual threat detection task," *Applied Cognitive Psychology*, 26 (2012), 194–206, and (2) P. Madhavan and J. R. Brown, "How (un)biased are airport security screening procedures? A social-cognitive experiment," *Current Research in Psychology*, 1(1)(2010), 71–74.
15 All statistics come from The Innocence Project, *DNA exonerations in the United States* (2019), www.innocenceproject.org/free-innocent/improve-the-law/fact-sheets/dna-exonerations-nationwide#sthash.Bc9uPlSK.dpuf (retrieved December 3, 2020).
16 J. Jones (November 25, 2019). Americans now support life in prison over the death penalty. Gallup News Service, https://news.gallup.com/poll/268514/americans-support-life-prison-death-penalty.aspx (retrieved September 23, 2021).
17 Also from *The Innocence Project* website, www.innocenceproject.org (retrieved December 3, 2020).
18 Ibid.
19 Elephants: B. L. Hart, L. A. Hart, and N. Pinter-Wollman, "Large brains and cognition: Where do elephants fit in?," *Neuroscience & Biobehavioral Reviews*, 32(1)(2008), 86–98. Rats: B. Vermaercke et al., "More complex brains are not always better: Rats outperform humans in implicit category-based generalization by implementing a similarity-based strategy," *Psychonomic Bulletin & Review*, 21(4)(2014), 1080–1086.

20 Quotes, narrative, and data were drawn from a number of informative articles about this project. In order of publication, they are: (1) J. V. Simmons, Jr., *Technical report 414: Project SEA HUNT FY 78 Final report* (San Diego, CA: Naval Ocean Systems Center, 1979), (2) R. E. Winter, "Move over, Lloyd Bridges: Project SEA Hunt," *Safe boating* (Washington, DC: Office of Boating Safety, United States Coast Guard, Department of Transportation, 1980), (3) J. V. Simmons, Jr., *Technical Report 746: Project Sea Hunt: A report on prototype development and tests* (San Diego, CA: Naval Ocean Systems Center, 1981), (4) L. Nash and H. G. Ketchen, *Evaluation of the detection capabilities of the SEA HUNT system* (Washington, DC: US Department of Transportation, United States Coast Guard, 1983).
21 A. Mahoney et al., "Mine detection rats: Effects of repeated extinction on detection accuracy," *Journal of Conventional Weapons Destruction*, 16(3)(2015), 61–64.
22 A. Poling et al., "Using trained pouched rats to detect land mines: Another victory for operant conditioning," *Journal of Applied Behavior Analysis*, 44(2)(2011), 351–355.
23 C. Rudder, *Dataclysm: Love, sex, race, and identity – What our online lives tell us about our offline selves* (New York: Broadway Books, 2014).
24 Two articles contributed to this claim: (1) R. A. Hubbard et al., "Cumulative probability of false-positive recall or biopsy recommendation after 10 years of screening mammography: A cohort study," *Annals of Internal Medicine*, 155(8) (2011), 481–492, and (2) E. D. Pisano et al., "Diagnostic accuracy of digital versus film mammography: Exploratory analysis of selected population subgroups in DMIST," *Radiology*, 246(2)(2008), 376–383.

Chapter 9

1 O. G. Selfridge and U. Neisser, "Pattern recognition by machine," *Scientific American*, 203(1960), 60–68.
2 I. Biederman, "Recognition-by-components: A theory of human image understanding," *Psychological Review*, 94(2)(1987), 115–147.
3 J. Wagemans et al., "A century of Gestalt psychology in visual perception: I. Perceptual grouping and figure–ground organization," *Psychological Bulletin*, 138(6)(2012), 1172–1217.
4 For a serious, scientific look at jizz, please see: R. Ellis, "Jizz and the joy of pattern recognition: Virtuosity, discipline and the agency of insight in UK naturalists' arts of seeing," *Social Studies of Science*, 41(6)(2011), 769–790. For photos of dozens of people all pointing binoculars in the same direction, see: J. Leo, "All that jizz: Don't look now, but birding is in," *Time Magazine* (May 25, 1987).
5 J. W. Tanaka and M. Taylor, "Object categories and expertise: Is the basic level in the eye of the beholder?," *Cognitive Psychology*, 23(3)(1991), 457–482.
6 Personal communication, January 15, 2021.
7 A. Alim-Marvasti et al., "Transient smartphone 'blindness,'" *The New England Journal of Medicine*, 374(25)(2016), 2502–2504.
8 S. A. Balk et al., "Highlighting human form and motion information enhances the conspicuity of pedestrians at night," *Perception*, 37(8)(2008), 1276–1284.
9 Centers for Disease Control (CDC), "Pedestrian safety. Centers for Disease Control and Prevention" (2021), www.cdc.gov/transportationsafety/pedestrian_safety/index.html (retrieved May 27, 2021).
10 Personal communication, January 15, 2021.
11 R. A. Tyrrell, J. M. Wood, and T. P. Carberry, "On-road measures of pedestrians' estimates of their own nighttime conspicuity," *Journal of Safety Research*, 35(5) (2004), 483–490.

12 The standard published by the Queensland Government, Australia, is called *Transport and Main Roads – Queensland Traffic Controller Clothing Standard 2016* and contains the text "From 1 July 2016 Class R tape strips to create a 'biomotion' (or 'biological motion') effect must be applied to all traffic controller shirts, weatherproof garments, and trousers worn by traffic controllers undertaking traffic control functions in periods of darkness."
13 V. S. Ramachandran, *The tell-tale brain* (New York: W.W. Norton & Company, 2010).
14 PBS, *Secrets of the mind* (October 23, 2001), NOVA.
15 S. Cahalan, *Brain on fire: My month of madness* (New York: Free Press, 2012).
16 J. Finney, *The body snatchers* (New York: Dell Publishing, 1955).
17 PBS, Secrets of the mind (October 23, 2001), NOVA.

Chapter 10

1 All quotes from LeClair came from: G. T. Morris, S. Spence, & M. Cryger, *Surgical items left in patients' bodies, USA Today* (video interview, March 7, 2013), www.usatoday.com/videos/news/health/2014/02/10/1972357/ (retrieved May 21, 2021).
2 N. N. Egorova et al., "Managing the prevention of retained surgical instruments," *Annals of Surgery*, 247(1)(2008), 13–18.
3 A. Gawande & T. Weiser (eds.), *WHO guidelines for safe surgery 2009: Safe surgery saves lives* (Geneva: Patient Safety, World Health Organization, 2009).
4 D. Fickling, "The cruel sea," *The Guardian* (July 22, 2004).
5 P. D'Oench, *Stranded divers rescued off Key Biscane*, CBS Local Miami (October 4, 2011), http://miami.cbslocal.com/2011/10/04/divers-rescue-after-charter-boat-leaves-without-them/ (retrieved May 21, 2021).
6 Associated Press, "Australian officials investigate after US snorkeller left behind in sea," *The Guardian* (June 29, 2011).
7 M. Roosevelt, "Forgotten at sea, scuba diver wins $1.68 million," *The Seattle Times* (October 24, 2010).
8 H. Chang & M. Harris, *Lost at sea: Expert diver mysteriously disappears*, NBC Los Angeles (February 20, 2017).
9 Scientific studies and reviews describing this physical nature of memory are readily available, namely: K. Nader, "Reconsolidation and the dynamic nature of memory," in *Novel mechanisms of memory* (Cham: Springer, 2016), pp. 1–20, and C. M. Alberini & J. E. LeDoux, "Memory reconsolidation," *Current Biology*, 23(17) (2013), R746–R750. For a well-explained and accessible overview, I recommend listening to the podcast episode: J. Abumrad & R. Krulwich, *Eternal sunshine of the spotless rat*, in *Memory and Forgetting* (2007). Radiolab.
10 The full story was retold, with analysis, in a *New York Times* op-ed: C. F. Chabris & D. J. Simons, "Why our memory fails us" (December 1, 2014).
11 Neil Degrasse Tyson posted this apology on Facebook, www.facebook.com/notes/10224657946775630/ (retrieved May 21, 2021).
12 A. Selk, "Falsely accused of satanic horrors, a couple spent 21 years in prison. Now they're owed millions," *Washington Post* (August 25, 2017).
13 D. Jenish, "End of a nightmare: Acquittals in Martensville cast doubt on how abuse cases are investigated," *Macleans* (February 14, 1994).
14 All quotes from the *Uncover* radio show came from: L. Bryn, *Uncover: Satanic panic* (CBC Radio-Canada, n.d.).
15 K. Lanning, *Investigator's guide to allegations of "ritual" child abuse* (Washington, DC: US Department of Justice, National Institute of Justice, 1992).

16 A. Spiegal, *An epidemic created by doctors*, in *Ask an Expert* (2002), *This American Life*.
17 D. A. Poole & D. S. Lindsay, "Interviewing preschoolers: Effects of nonsuggestive techniques, parental coaching, and leading questions on reports of nonexperienced events," *Journal of Experimental Child Psychology*, 60(1)(1995), 129–154.
18 C. Newlin et al., *Child forensic interviewing: Best practices* (Washington, DC: Office of Juvenile Justice and Delinquency Prevention. US Department of Justice, 2015).
19 American Psychological Association, "Questions and answers about memories of childhood abuse" (1995), www.apa.org/topics/trauma/memories (retrieved May 21, 2021).
20 Ibid.
21 A. Bellos, "He ate all the pi: Japanese man memorises π to 111,700 digits," *The Guardian* (March 2015), www.theguardian.com/science/alexs-adventures-in-numberland/2015/mar/13/pi-day-2015-memory-memorisation-world-record-japanese-akira-haraguchi (retrieved October 1, 2020).
22 Ibid.
23 Statistics available in these reviews: D. Hariharan & D. N. Lobo, "Retained surgical sponges, needles and instruments," *The Annals of the Royal College of Surgeons of England*, 95(2)(2013), 87–92. http://doi.org/10.1308/003588413X13511609957218, and A. A. Gawande et al., "Risk factors for retained instruments and sponges after surgery," *New England Journal of Medicine*, 348(3)(2003), 229–235. http://doi.org/10.1056/nejmsa021721
24 R. R. Cima et al., "Using a data-matrix-coded sponge counting system across a surgical practice: Impact after 18 months," *The Joint Commission Journal on Quality and Patient Safety*, 37(2)(2011), 51-AP3. http://doi.org/10.1016/s1553-7250(11)37007-9.
25 CDC, *Contact tracing resources for health departments: Resources for conducting contact tracing to stop the spread of covid-19* (2020), www.cdc.gov/coronavirus/2019-ncov/php/open-america/contact-tracing-resources.html (retrieved October 1, 2020). CDC, *Key information to collect during a case interview* (2020), www.cdc.gov/coronavirus/2019-ncov/php/contact-tracing/keyinfo.html (retrieved October 1, 2020).
26 Ibid.

Chapter 11

1 D. J. Gillan, "Five questions concerning task analysis," in M. A. Wilson, W. Bennett Jr., S. G. Gibson, & G. M. Alliger (eds.), *The handbook of work analysis: The methods, systems, applications, & science of work measurement in organizations* (New York: Routledge), pp. 201–214.
2 B. Crandall et al., *Working minds: A practitioner's guide to cognitive task analysis* (Cambridge, MA: MIT Press, 2006).
3 F. Isikdogan, T. Gerasimow, & G. Michael, "Eye contact correction using deep neural networks," in *The IEEE Winter Conference on Applications of Computer Vision* (Snowmass Village, CO: IEEE, 2020), pp. 3318–3326.
4 All quotes and information from Laura Levy came from an interview I conducted with her in December, 2020.
5 A. Perrin, *5 facts about Americans and video games*, Pew Research Center (2018).
6 L. Lanier, "Video games could be a $300 billion dollar industry by 2025," *Variety Magazine* (May 1, 2019).

7 S. Parkin, "Desert bus: The very worst video game ever created," *The New Yorker* (September 29, 2013).
8 Publications on the topic include C. Hodent, *The gamer's brain: How neuroscience and UX can impact video game design* (Boca Raton, FL: CRC Press, 2017), and C. Hodent, *The psychology of video games* (Abingdon: Routledge, 2020).
9 T. Nichols, *An interesting career in psychological science: Video game user researcher* (Psychological Science Agenda, APA, 2011), www.apa.org/science/about/psa/2009/11/careers (retrieved December 5, 2020).
10 E. Maiberg, "'Minecraft' VR is the best game on the oculus rift, but that's not saying much," *Vice Media* (2016), www.vice.com/en/article/qkjwy7/minecraft-vr-review (retrieved November 28, 2020).
11 J. Porter III, M. Boyer, & A. Robb, "Guidelines on successfully porting non-immersive games to virtual reality: A case study in Minecraft," in *Proceedings of the 2018 Annual Symposium on Computer–Human Interaction in Play* (New York: Association for Computing Machinery, 2018), pp. 405–415.
12 Personal communication with Celia Hodent, December 20, 2020.

REFERENCES

Behavioral Science Technology (2004). *Assessment and plan for organizational culture change at NASA*. Retrieved June 30, 2020 from www.nasa.gov/pdf/57382main_culture_web.pdf.

Biggs, A. T. & Mitroff, S. R. (2015). Improving the efficacy of security screening tasks: A review of visual search challenges and ways to mitigate their adverse effects. *Applied Cognitive Psychology*, 29(1), 142–148.

Chabris, C. F. & Simons, D. (2010). *The invisible gorilla: And other ways our intuitions deceive us*. New York: Crown.

Clark, S. E. (2012). Costs and benefits of eyewitness identification reform: Psychological science and public policy. *Perspectives on Psychological Science*, 7(3), 238–259. doi: 10.1177/1745691612439584.

Dixon, S. R., Wickens, C. D., & McCarley, J. S. (2007). On the independence of compliance and reliance: Are automation false alarms worse than misses? *Human Factors*, 49(4), 564–572. doi: 10.1518/001872007X215656.

Edewaard, D. E., Fekety, D. K., Szubski, E. C., & Tyrrell, R. A. (2020). Highlighting bicyclist biological motion enhances their conspicuity in daylight. *Accident Analysis & Prevention*, 142, 105575.

Endsley, M. A. & Kiris, E. O. (1995). The out-of-the-loop performance problem and level of control in automation. *Human Factors*, 37(2), 381–394.

The Independent Investigation on the Fukushima Nuclear Accident (2014). *The Fukushima Daiichi nuclear power station disaster*. London: Routledge.

International Atomic Energy Agency (1992). The Chernobyl accident: Updating of INSAG-1. IAEA. Retrieved February 11, 2020 from www-pub.iaea.org/MTCD/publications/PDF/Pub913e_web.pdf.

Kohn, L. T., Corrigan, J. M., & Donaldson , M. S. (2000). To err is human: Building a safer health System. In *Institute of Medicine (US) Committee on quality of health care in America*. Washington, DC: National Academy Press.

Lewis, M. B., Sherwood, S., Moselhy, H., & Ellis, H. D. (2001). Autonomic responses to familiar faces without autonomic responses to familiar voices: Evidence for voice-specific Capgras delusion. *Cognitive Neuropsychiatry*, 6(3), 217–228.

Mantyla, T., Still, J., Gullberg, S., & Del Missier, F. (2014). Decision making in adults with ADHD. *Journal of Attention Disorders*, 14(2), 166–173.

McCall, C. (2016). Chernobyl disaster 30 years on: Lessons not learned. *The Lancet*, 387(10029), 1707.

Motosaka, M. & Mitsuji, K. (2012). Building damage during the 2011 off the Pacific coast of Tohoku earthquake. *Soils and Foundations*, 52(5), 929–944.

Navon, D. & Gopher, D. (1979). On the economy of the human-processing system. *Psychological Review*, 86(3), 214–255. http://dx.doi.org/10.1037/0033-295X.86.3.214.

Norris, P. & Epstein, S. (2011). An experiential thinking style: Its facets and relations with objective and subjective criterion measures. *Journal of Personality*, 79(5), 1043–1080.

Paivio, A. (1986). *Mental representations: A dual coding approach*. New York: Oxford University Press

Reason, J. (1995). A system approach to organizational error. *Ergonomics*, 38, 1708–1721. doi: 10.1080/00140139508925221.

Speed, A., Bonnie, K., & Caggiano, D. (2015). *Human factors within the transportation security administration: Optimizing performance through human factors assessments* (No. SAND2015-1563C) (Albuquerque, NM: Sandia National Lab.[SNL-NM]).

Tokimatsu, K., Tamura, S., Suzuki, H., & Katsumata, K. (2012). Special issue on geotechnical aspects of the 2011 off the Pacific coast of Tohoku earthquake. *Soils and Foundations*, 52(5), 929–944.

Tyrrell, R. A., Wood, J. M., Chaparro, A., Carberry, T. P., Chu, B.-S., & Marszalek, R. P. (2009). Seeing pedestrians at night: Visual clutter does not mask biological motion. *Accident Analysis & Prevention*, 41, 506–512.

Tyrrell, R. A., Wood, J. M., Owens, D. A., Whetsel Borzendowski, S., & Stafford Sewall, A. (2016). The conspicuity of pedestrians at night: A review. *Clinical and Experimental Optometry*, 99(5), 425–434.

Unknown (1962). Petrosyants appointed head of Soviet Atomic Energy Committee. *Open Society Archives, Central European University*. Retrieved June 6, 2016 from http://w3.osaarchivum.org/files/access/rip/10/300-80-1-55-000157.pdf.

US Senate & United States Senate Armed Services Committee (2008). Inquiry into the treatment of detainees in US custody. In *Report of the Committee on Armed Services United States Senate, 110th Congress, 2nd Session (Vol. 20)*.

Wickens, C. D. & Carswell, C. M. (1995). The proximity compatibility principle: Its psychological foundation and relevance to display design. *Human Factors*, 37, 473–494.

Wood, J. M., Marszalek, R., Lacherez, P., & Tyrrell, R. A. (2014). Configuring reflective markings to enhance the nighttime conspicuity of road workers. *Accident Analysis & Prevention*, 70, 209–214.

Wood, J. M., Tyrrell, R. A., & Carberry, T. P. (2005). Limitations in drivers' ability to recognize pedestrians at night. *Human Factors*, 47(3), 644–653.

Wood, J. M., Tyrrell, R. A., Marszalek, R., Lacherez, P., Carberry, T. P., & Chu, B. S. (2012). Using reflective clothing to enhance the conspicuity of bicyclists at night. *Accident Analysis & Prevention*, 45, 726–730.

Wood, J. M., Tyrrell, R. A., Marszalek, R., Lacherez, P., Chaparro, A., & Britt, T.W. (2011). Using biological motion to enhance the conspicuity of roadway workers. *Accident Analysis & Prevention*, 43, 1036–1041.

INDEX

CPSIA information can be obtained
at www.ICGtesting.com
Printed in the USA
LVHW101934301022
731935LV00005B/134

9 781009 012546